Herb Watson

Fundamentals of Small Gas Engines

FUNDAMENTALS OF SMALL GAS ENGINES

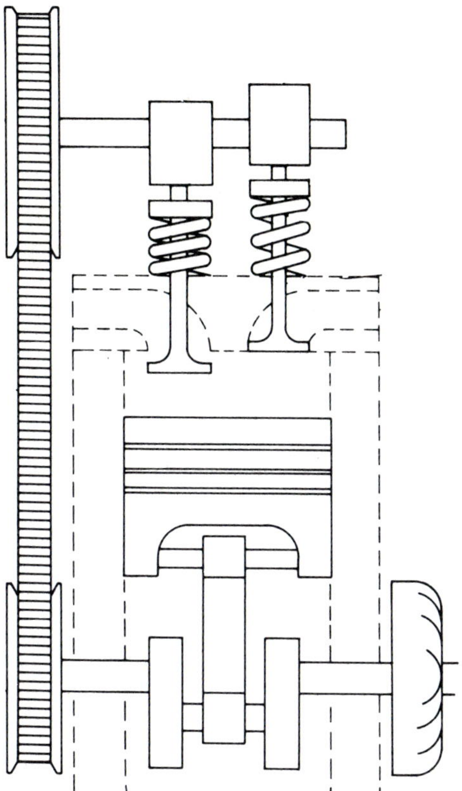

Tom Weathers Jr.

Claud Hunter

Reston Publishing Company, Inc.
A Prentice-Hall Company
Reston, Virginia

Library of Congress Cataloging in Publication Data

Weathers, Tom.
Fundamentals of small gas engines.

Includes index.
1. Internal combustion engines, Spark ignition.
I. Hunter, Claud C. II. Title.
TJ790.W388 1983 621.43'4 83-6211
ISBN 0-8359-2192-1
ISBN 0-8359-2191-3 (pbk.)

10 9 8 7 6 5 4 3 2

Printed in the United States of America.

CONTENTS

PREFACE

Many of the changes taking place in the twentieth century have been credited to various advances in technology:

1. The automobile altered the way we view distance and travel.
2. Television gave us a new perspective on information.
3. The computer changed the way we look at ourselves as thinking creatures.

These inventions (plus, of course, electricity, airplanes, atom bombs, and a few others) are generally regarded as having the most effect on our lives. However, there is another level of everyday technology that is so common we seldom think about it. Refrigerators, electric stoves, and air conditioners are typical representatives of this category, as is the small gasoline engine, the subject of this book.

Consider how many ordinary tasks probably wouldn't even be attempted without the assistance of a small engine. Mowing and maintaining a large, grassy lawn is one example. Cutting wood for a fireplace is another. The list could go on and on to include both applications on the job and around the house. Because of its versatility and utility, the small gasoline engine has become a significant replacement for human muscle power.

This book is for anyone who wants to become familiar with the basic operation and components of small gasoline engines. No prior knowledge is assumed. Starting with the basics, information is added in a building-block manner. Briefly, here is what each chapter covers.

CHAPTER 1 gives the background of internal combustion engines and defines the term "small" as it relates to small gasoline engines.

CHAPTER 2 introduces the main operating systems and explains two- and four-stroke cycle operation.

CHAPTER 3 looks at some common ways to compare the size and performance of small engines.

CHAPTER 4 examines the design and construction of the major components used in small engines.

CHAPTER 5 reviews the major types of fuels used in small engines and examines some typical small engine fuel systems.

CHAPTER 6 introduces the fundamentals of electricity and magnetism.

CHAPTER 7 examines some typical small engine ignition systems.

CHAPTER 8 looks at manual and electrical means used to crank small engines as well as electrical generating systems employed by some small engines.

CHAPTER 9 reviews lubrication system fundamentals and then examines typical small engine lubrication systems.

CHAPTER 10 looks at small engine cooling and exhaust systems.

CHAPTER 11 examines small engine valve train operation and service.

CHAPTER 12 examines small engine piston and connecting rod service.

CHAPTER 13 examines engine block, cylinder, and crankshaft service for small engines.

CHAPTER 14 reviews routine maintenance and tune up procedures for small engines.

CHAPTER 15 introduces common hand tools used to service small engines.

CHAPTER 16 looks at various types of measuring tools.

The information contained in these chapters will not make you expert on small engines. However, it will provide a framework on which you can hang the facts that you learn later on your own. Add enough facts from practical experience to a good, solid framework, and you will become an expert.

1

INTRODUCTION TO SMALL ENGINES

GENERAL

This chapter discusses the definition of small gasoline engines. (What is small?) It also notes some general uses for small engines and gives a few brief facts about the development of internal combustion engines.

BASIC QUESTIONS

Before we get into the development story, it will be helpful to answer a few basic questions.

First of all, what does the term "internal combustion" mean? *Answer*: It refers to the fuel burning process that takes place inside an engine. (See Figure 1-1.) When the fuel burns, chemical energy is released in the form of hot, expanding gases. These gases push on a piston which turns a crankshaft, which drives whatever the engine is attached to. Next, you might ask, what is chemical energy? *Answer*: For our purposes, we can say that chemical energy is a fuel's ability to do work. It is as though a fuel, such as gasoline, is filled with tiny, tightly wound springs. When enough heat is present, the springs unwind to release their energy.

BACKGROUND OF INTERNAL COMBUSTION ENGINES

Developers of the first internal combustion engines were also concerned with these basic questions. However, without the benefits of modern technology, the results of their work hardly resembled the engines we know today.

The first attempt to create an internal combustion engine probably

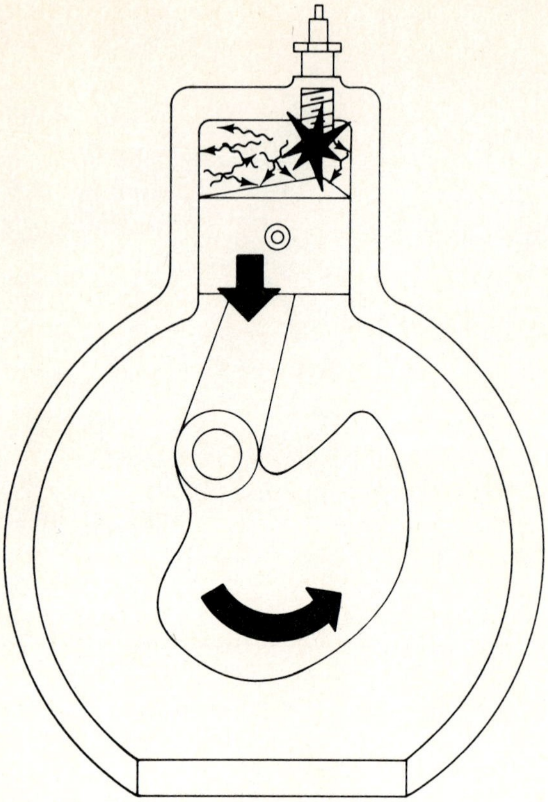

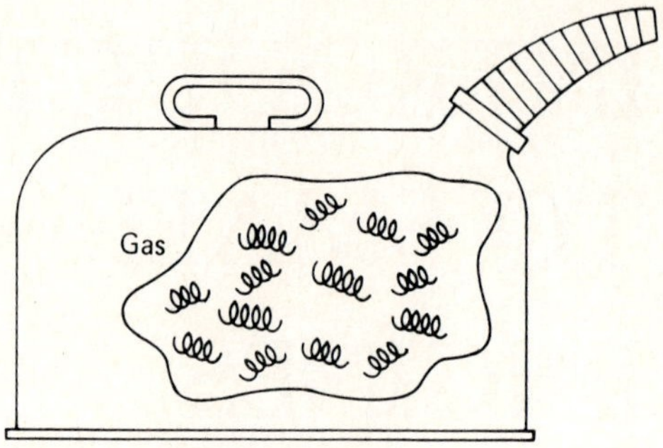

Chemical Energy is like Tightly Coiled Springs which Unwind when Fuel is Ignited.

FIGURE 1-1. Internal combustion releases the stored energy in fuel.

took place in the late 1600s. French inventors tried to find a controlled way to release the explosive energy contained in gunpowder. After all, they may have reasoned, if gunpowder could move a projectile, it ought to be able to operate an engine. Unfortunately the idea didn't work. All they created was a new, very dangerous cannon.

Interestingly, the young Thomas Edison, according to one historian, also considered gunpowder as a fuel for internal combustion engines. However, like his French predecessors, he also decided that it wasn't a very good idea.

The first successful work on internal combustion engines occurred around the middle of the 1800s. It was during this period that another Frenchman, named de Rochas, formulated basic internal combustion theories and a German, Dr. Nicholas Otto, developed the four-stroke principle of operation. Even now, four-stroke engines are sometimes referred to as Otto cycle engines.

Early experimenters attempted to use various kinds of fuel. One of the most common fuels was the "illuminating gas" normally employed in the gas-fired lamps of the time. It was made from coal or heavy oils. The first operational gasoline-powered engine is reported to have been constructed by still another Frenchman, Renoir, around 1870.

None of these early engines were physically small; they weighed 800 to 1200 pounds for each horsepower produced. On the other hand, almost all were small in terms of output and rotational speed. In fact, some of the early engines did not rotate at all. They were designed to operate large pumps and depended solely on a back and forth, reciprocating action.

Crankshafts were not employed to change the reciprocating motion into force applied in a circular manner (e.g., torque).

DEVELOPMENT OF SMALL GASOLINE ENGINES

It is difficult to say exactly when the first "small" gasoline engine appeared. That is because it is difficult to define exactly what we mean by a small gasoline engine.

If we view strictly from the standpoint of size, we might say that the first small gasoline engine was one that an average person could pick up and carry around. This state of development probably occurred around the end of the nineteenth century.

However, defining small gasoline engines strictly in terms of size doesn't seem quite right. Small automobile engines, even those lightweight enough to be carried around, are not usually grouped into the small engine category. The same distinction holds true for marine, airplane, and motorcycle engines. Although sometimes physically small, these engines are most often included in their own separate categories.

Small engines seem to fall into a leftover category not occupied by other types of engines. So, we could say that the first small gasoline engine was the first lightweight engine not used in an automobile, boat, airplane, or motorcycle. (See Figure 1-2.) Just how was this engine used? It was likely employed in a somewhat personal fashion as a direct replacement for human muscle power. It could have been used to operate a grass mower or wood cutting saw. Or the first small engine may have provided the motive force for a device that would have been otherwise impractical without a small, lightweight powerplant. Water pumps, small saw mills, and portable

FIGURE 1-2. Early small engine (Courtesy of Onan).

electric generators are some possibilities. Gasoline engines used in early washing machines are another possibility.

CURRENT VIEW OF SMALL GASOLINE ENGINES

No matter how they are defined or when they originated, small engines have had a remarkable effect on the average person. By enhancing our muscle power, small engines let us do work our grandfathers wouldn't even dream of. Can you imagine mowing a 200 by 200-ft grass yard with a reel-type push mower? It is only because of gasoline-powered lawn mowers that large, grassy lawns have become common. No one would have the time to tend them otherwise. What about your new wood-burning stove? Although your grandfather may have cut wood with an axe and saw, you probably don't have the time. A small engine-powered chain saw makes the proposition much more practical. The list goes on and on – not just for private use, but for commercial and industrial applications as well. (See Figure 1-3.)

In this book, we will define small engines as having between one and four cylinders and producing up to 20 or 30 horsepower. Although a number of motorcycle and marine engines fit this definition, as we have noted before, they are usually grouped into their own separate categories. We will concentrate on the general-purpose small engines, those that don't fit into the other categories. We will view the small gasoline engine as a tool – versatile enough to perform an almost endless variety of jobs.

FIGURE 1-3. Modern small engines (Courtesy of Onan).

2

FUNDAMENTALS OF OPERATION

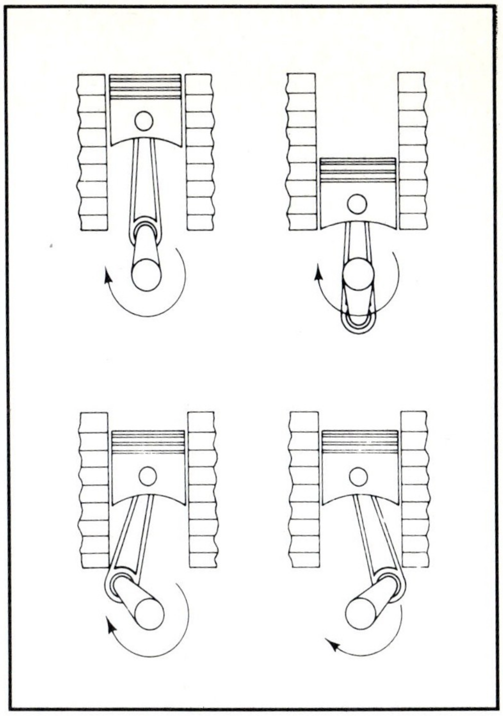

GENERAL

This chapter examines the fundamental operating principles of small gasoline engines. The combustion process is introduced; basic operating systems are reviewed, and four- and two-stroke cycles of operation are explained.

COMBUSTION PROCESS

As noted in the first chapter, a fuel, such as gasoline, contains chemical energy—acting like tiny springs waiting to pop loose. However, fuel can't release the energy by itself. Air and heat are needed. The air gets the energy ready to be released, in effect stretching the tiny springs. Heat acts like a trigger to set the energy free. Once released, the energy is converted into mechanical force to drive the engine and whatever it is attached to.

Fuel, heat, and air are the basic ingredients of the combustion process. The purpose of any engine, large or small, is to produce combustion and then to manage the results.

INTRODUCTION TO MAIN OPERATING SYSTEMS

In order to bring fuel, air, and heat together, all engines contain a number of basic operating systems. The next paragraphs introduce these systems; the following chapters describe the systems in detail.

Mechanical System

The components in the mechanical system are responsible for capturing the fuel's released energy and converting it into useful work. The mechanical system is also responsible for moving air and fuel through the engine and removing the exhaust gas.

The largest component in the mechanical system is the *block*. (See Figure 2-1.) It is a metal casting made out of iron or aluminum. Within the block are one or more openings called *cylinders*. Fuel and air are delivered inside the cylinders to be burned.

The bottom part of each cylinder is sealed by a sliding plug called the *piston*. The piston is connected by a rod to the crankshaft. The expanding gases created when the air and fuel are burned push down on the top of the piston. As the piston slides down in the cylinder, it rotates the crankshaft.

The crankshaft is usually connected to a flywheel, which, in turn, is usually attached to a clutch. Force created by the expanding gas of the combustion process is transmitted in this manner—to the piston, piston rod, crankshaft, flywheel, and clutch—into whatever the engine is driving.

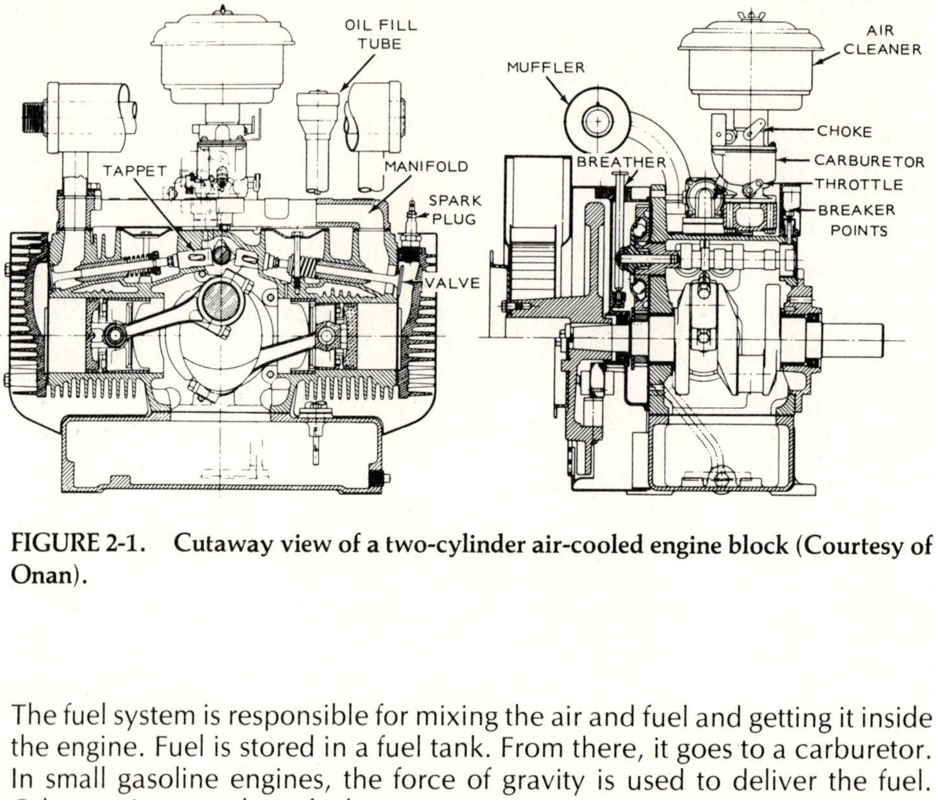

FIGURE 2-1. Cutaway view of a two-cylinder air-cooled engine block (Courtesy of Onan).

Fuel System

The fuel system is responsible for mixing the air and fuel and getting it inside the engine. Fuel is stored in a fuel tank. From there, it goes to a carburetor. In small gasoline engines, the force of gravity is used to deliver the fuel. Other engines employ a fuel pump.

The main job of the carburetor is breaking the fuel into tiny droplets and mixing the droplets with air (which has been cleaned by an air cleaner). The carburetor in most small engines also contains a choke for cold starting and a throttle to vary the amount of air and fuel delivered to the engine. This controls the speed of the engine.

After being mixed, the air and fuel go into the cylinder through an opening called the *intake port*. A valve, called the *intake valve*, controls the flow through the opening. When the valve is open, the air and fuel are drawn inside the cylinder to be burned. (See Figure 2-2.)

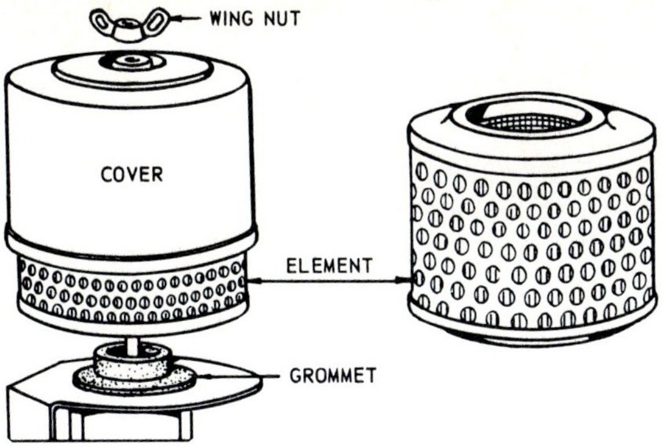

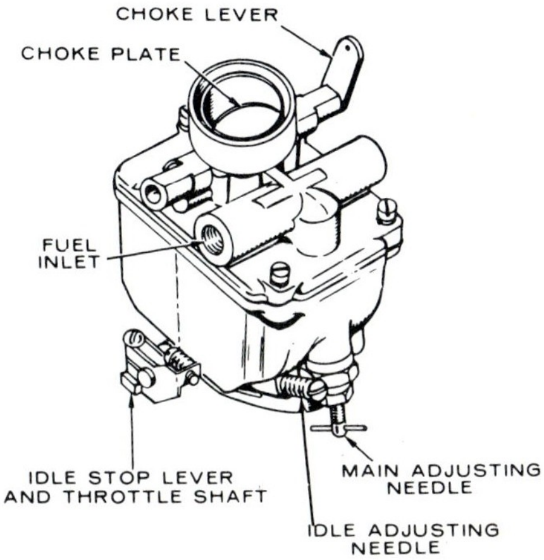

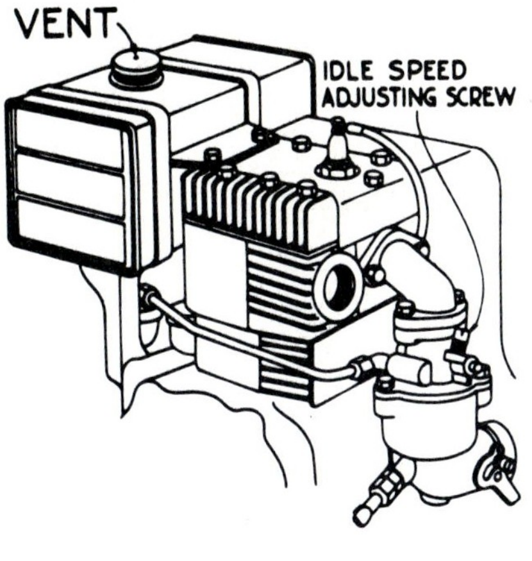

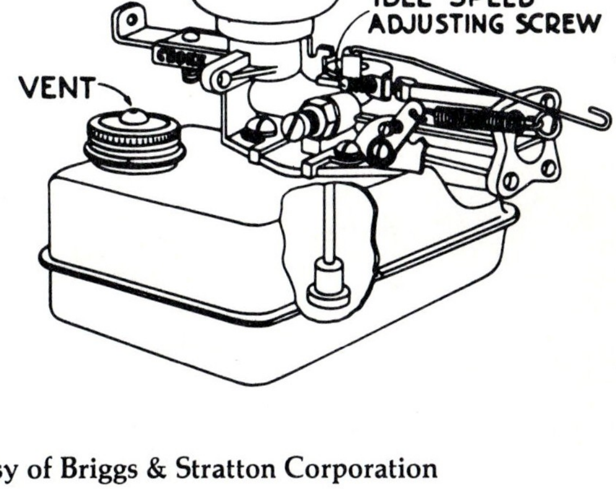

FIGURE 2-2. Fuel system components (Courtesy of Briggs & Stratton Corporation and Onan).

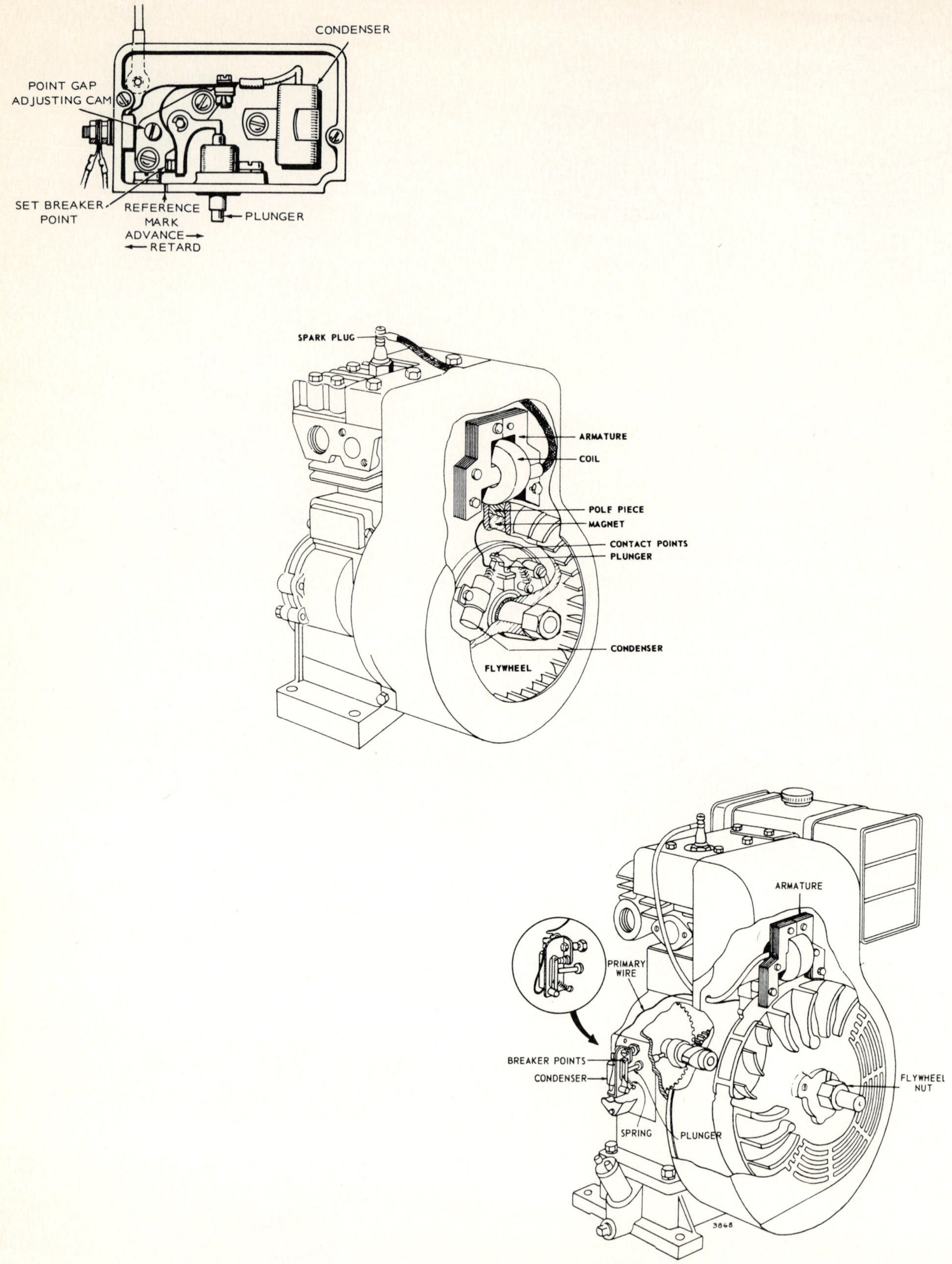

FIGURE 2-3. Ignition system components (Courtesy of Briggs & Stratton Corporation and Onan).

Ignition System

The main purpose of the ignition system is to supply the heat needed to start the combustion process. The heat occurs in the form of an electric spark which appears at the end of a spark plug projecting into the top of the cylinder. Electricity for the spark plug comes from the ignition coil. It converts low-voltage current from the engine's electrical system into momentary high-voltage surges.

As you'll see in Chapter 6, the coil supplies these high-voltage surges whenever the low-voltage circuit is interrupted. Many small engines use a mechanical switch called the *points*, to interrupt the low-voltage circuit. This action is timed to occur at just the right moment for proper combustion, when the piston is near the top of the cylinder. (See Figure 2-3.)

Exhaust System

Not all the fuel is consumed in the combustion process. The leftover byproducts, or exhaust, are passed out of another opening in the cylinder called the *exhaust port*. A valve (this time called the *exhaust valve*) controls the flow through the passage. When the valve is open, the exhaust goes out of the cylinder, into an exhaust manifold and from there into a muffler. (See Figure 2-4.) The muffler not only quietens the noise of the engine, it also helps catch sparks. This is especially important when small engines are being used in a forest, or otherwise near combustible materials.

FIGURE 2-4. Typical two-cylinder exhaust manifold and muffler assembly (Courtesy of Onan).

Lubricating System

The lower part of the block, called the *crankcase*, provides a reservoir for lubricating oil in most small engines. The oil, which is splashed or pumped to the moving surfaces in the engine, serves several purposes:

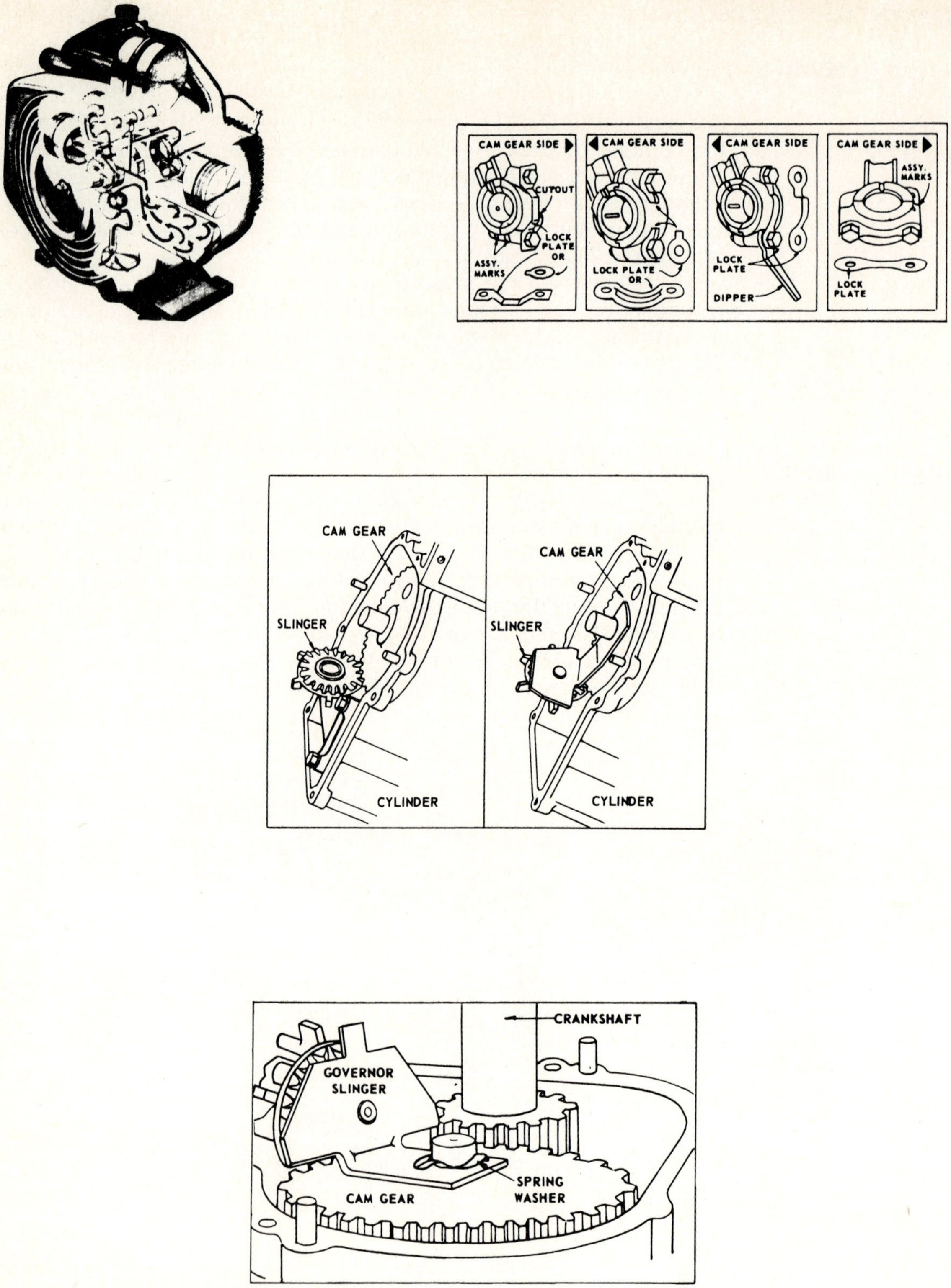

FIGURE 2-5. Oil flow and filtration system for a two-cylinder air-cooled engine (Courtesy of Briggs & Stratton Corporation and Onan).

1. *Lubrication.* Oil prevents moving or sliding surfaces from coming into direct contact with one another. This helps the engine run more efficiently and last longer.
2. *Cooling.* Heat from hot metal surfaces in the engine is transferred to the oil. The oil, in turn, transfers the heat to the bottom of the crankcase, which then radiates the heat to the atmosphere.
3. *Cleaning.* Oil washes away small particles and debris that are created when the engine is running. These waste products, which are usually heavier than the oil, settle to the bottom of the crankcase. (See Figure 2-5.)

Cooling System

Not only is heat required to start the combustion process, heat is also produced as a result of combustion. Getting rid of this excess heat is the job of the cooling system. (See Figure 2-6.) Some small engines transfer the heat to a cooling fluid which circulates through the block. The heat contained in the fluid is then transmitted by a radiator into the atmosphere. However, most small engines transfer the heat directly from the engine surfaces to the air. A fan is sometimes provided to make sure that enough cooling air flows over the engine. Fins cast into the block and other engine components are also used to get rid of excess heat. The additional surface area provided by the fins lets the heat escape more easily.

FIGURE 2-6. Cooling system.

Electrical System

All engines need a source of electrical energy. In very small engines, electricity is needed only for the ignition system. Larger engines may require electricity to operate a starter motor, lights, a horn, etc.

Electricity is produced by using the engine's output to turn a generating device operating off the end of the crankshaft. Larger engines may employ a battery and generator. Smaller engines usually only have a generator, called in these applications a *magneto*. (See Figure 2-7.)

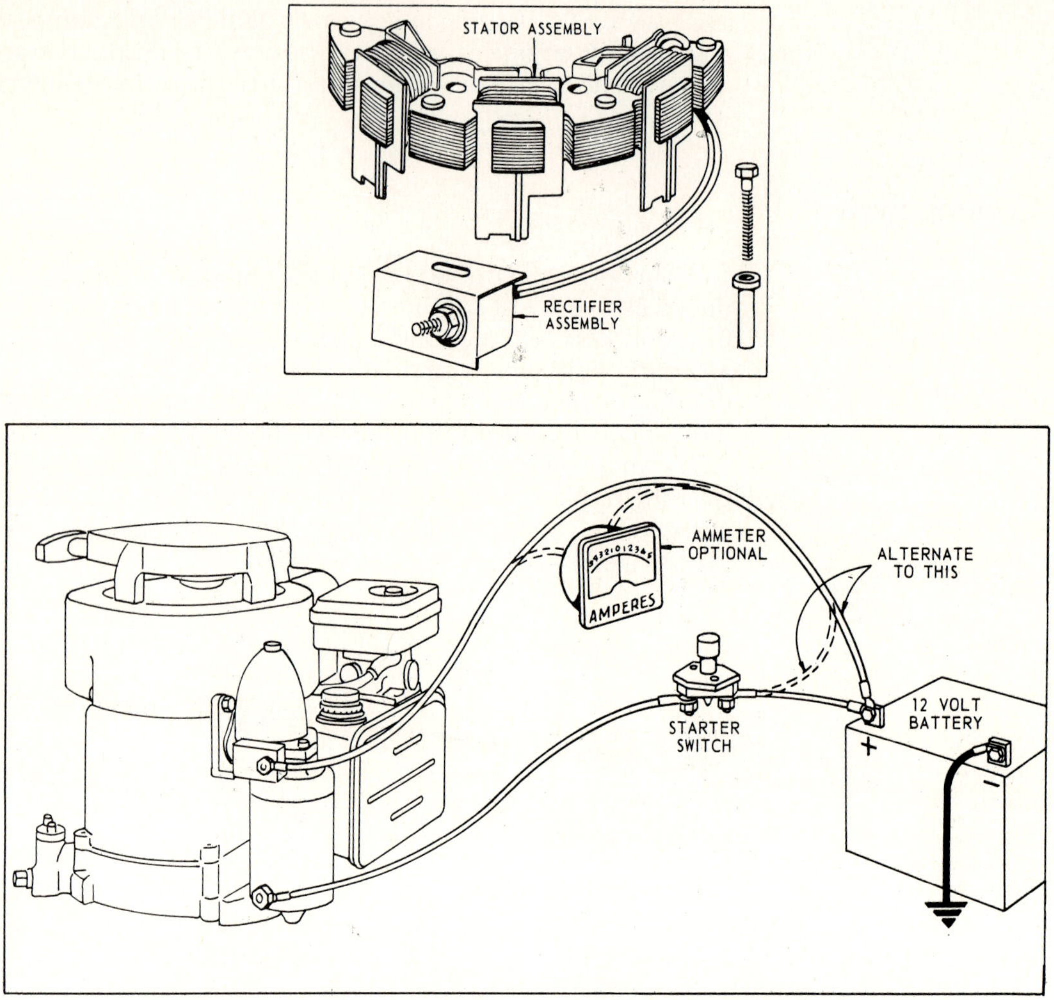

FIGURE 2-7. Flywheel alternator system (Courtesy of Briggs & Stratton Corporation).

MECHANICAL SYSTEM IN MORE DETAIL

Before we look at four- and two-stroke operating cycles, it will be helpful to re-examine the mechanical system. (These components will be examined in even more detail in Chapter 4.)

Block/Cylinder/Head

As we noted before, the block is the metal casting that contains the cylinder(s). [See Figure 2-8(a).] In principle, a cylinder is simply a round, smooth, metal tube. [See Figure 2-8(b).] In some engines, the cylinder is

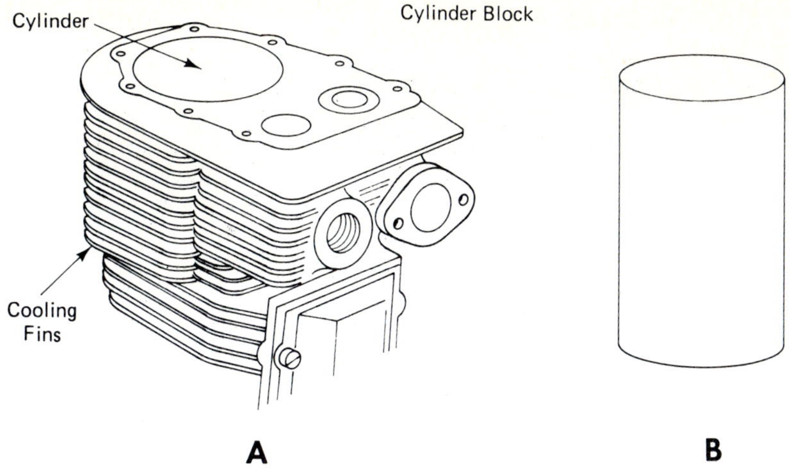

FIGURE 2-8. (A) Upper part of cylinder block showing top of cylinder; (B) cylinder (Courtesy of Briggs & Stratton Corporation).

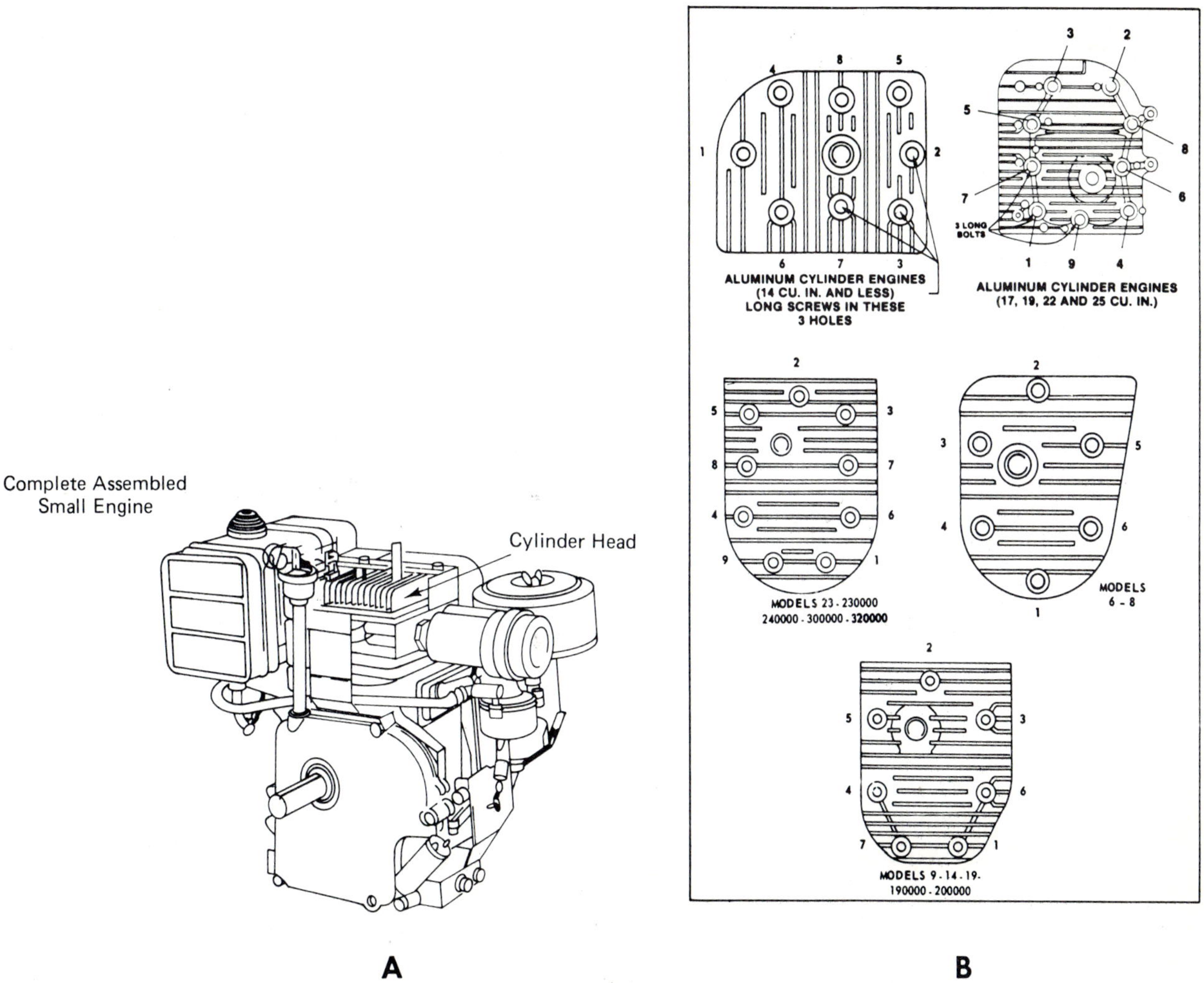

FIGURE 2-9. (A) Complete engine with cylinder heads indicated; (B) cylinder heads (Courtesy of Briggs & Stratton Corporation).

bored directly into the block material; in other engines, the cylinder actually is a tube pressed into a larger opening in the block.

The cylinder head is like a lid or cover that seals off the top of the cylinder. Sometimes the head is cast integrally with the block. In most engines, however, the head is a separate casting attached to the block with cap screws. [See Figure 2-9.] Sealing between the head block is provided by a special cylinder head gasket.

Piston/Rings

The piston, as indicated previously, is a movable plug that seals the end of the cylinder, opposite from the head. The piston fits closely into the cylinder, but not so closely as to inhibit movement. The sliding seal between the piston and cylinder walls is improved by the piston rings. They fit into grooves machined around the top or head of the piston. (See Figure 2-10.)

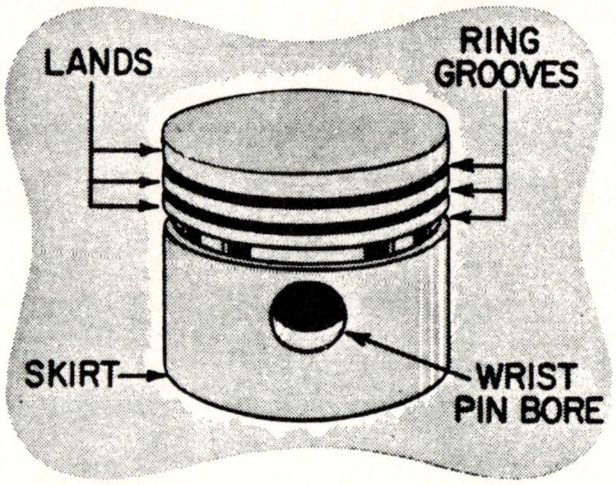

FIGURE 2-10. Diagram of a piston showing ring grooves, lands, wrist pin bore, and skirt (Courtesy of Clinton Engines Corporation).

Piston Pins/Rods

A hole is passed from one side of the piston to another, just below the ring area. The piston pin, or *wrist pin*, as it is sometimes called, fits into the hole. The pin acts as a pivot point for the piston rod which connects the piston to the crankshaft. (See Figures 2-11 and 2-12.)

The rod (see Figure 2-13) serves two purposes. It transmits force from the piston to the crankshaft. It also helps to convert the piston's up and down movement into the circular motion of the crankshaft. The pedal and

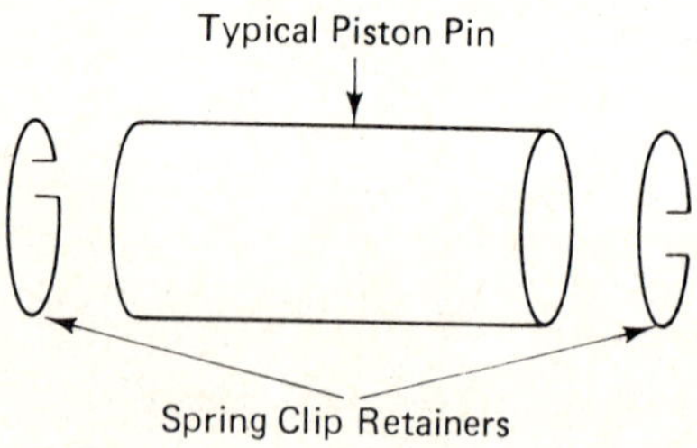

FIGURE 2-11. A typical piston pin.

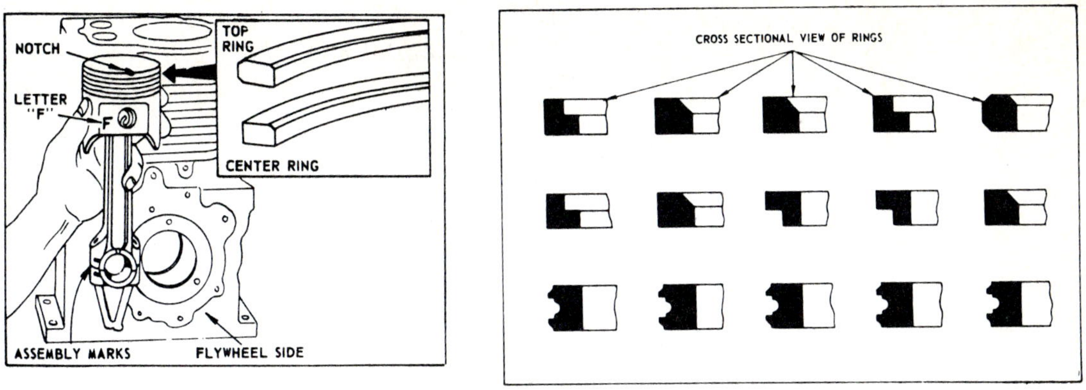

FIGURE 2-12. Piston, rod, and rings (Courtesy of Briggs & Stratton Corporation).

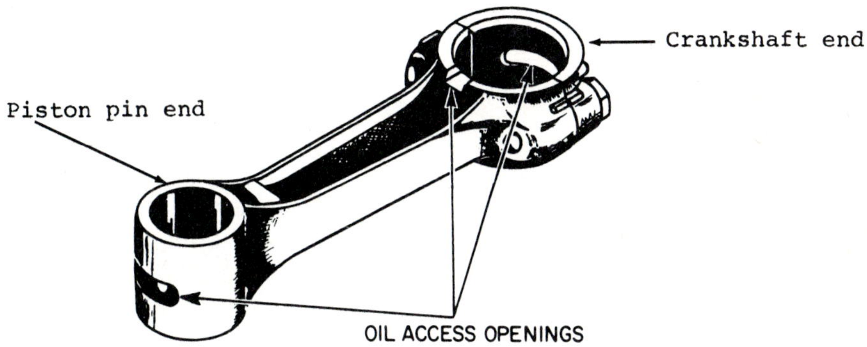

FIGURE 2-13. A connecting rod (Courtesy of Clinton Engines Corporation).

crank assembly of a bicycle works the same way to convert up and down leg movements into circular wheel motion.

To ensure smooth application of force, the rod is attached to bushings or bearings at the upper end and a bearing surface on the crankshaft at the lower end. The crankshaft itself is supported at both ends by main bearings or bushings which are fitted into "bosses" machined into the block. (See Figure 2-14.)

Valve Train

As noted before, two openings, the intake and exhaust ports, lead into the cylinders. (See Figure 2-15.) These openings, which are usually located in the head, are used for incoming air and fuel and outgoing exhaust gases. Valves, technically known as *poppet valves*, act as doors for the ports. (See Figure 2-16.)

The position of the valves (open or closed) is controlled by a cam acting through a series of rods and linkages known collectively as the *valve train*. The cam acts like a rotating inclined plane to push the valves open at just the right time. Springs close the valves when the cam rotates to a low point.

The cam surfaces or lobes are located on a camshaft. (See Figure 2-17.) There is one lobe for each intake and exhaust valve. A gear, permanently mounted on the end of the camshaft, is driven by another gear

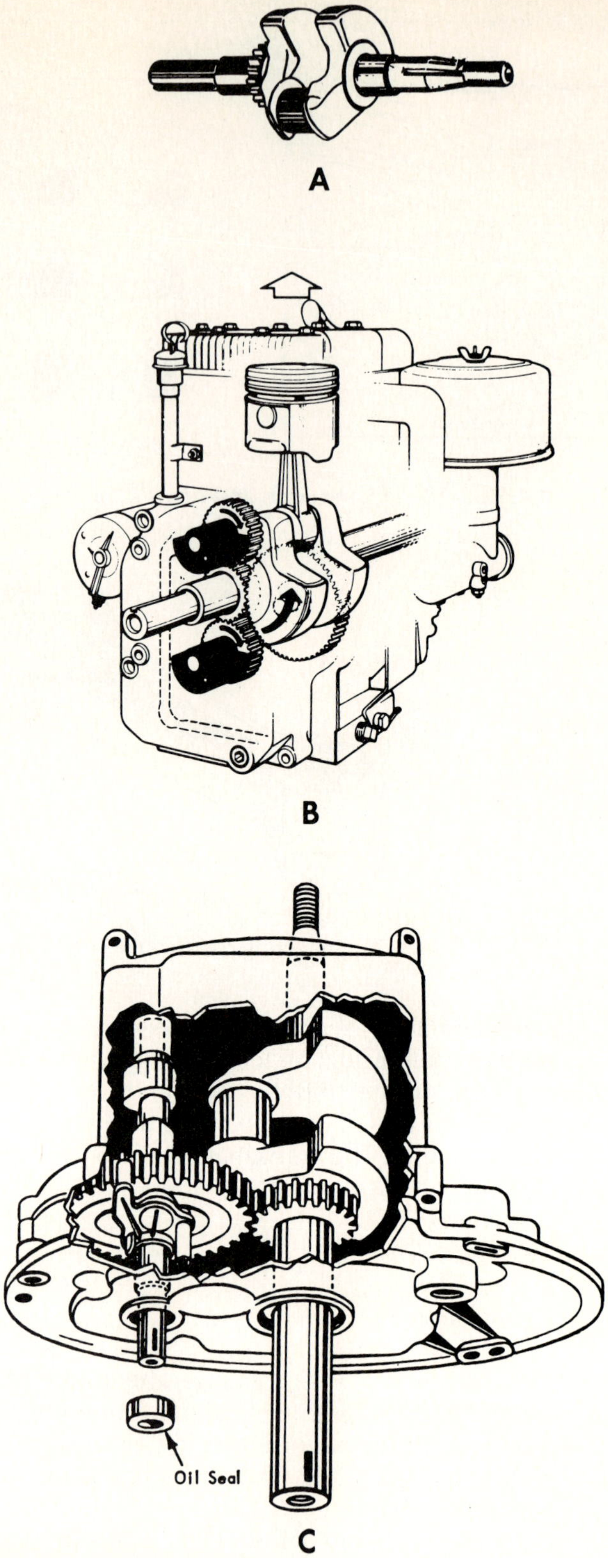

FIGURE 2-14. (A) Typical crankshaft for a single-cylinder small engine; (B) piston, connecting rod, crankshaft, and crankcase; (C) cam gear relations (Courtesy of Clinton Engines Corporation and Briggs & Stratton Corporation).

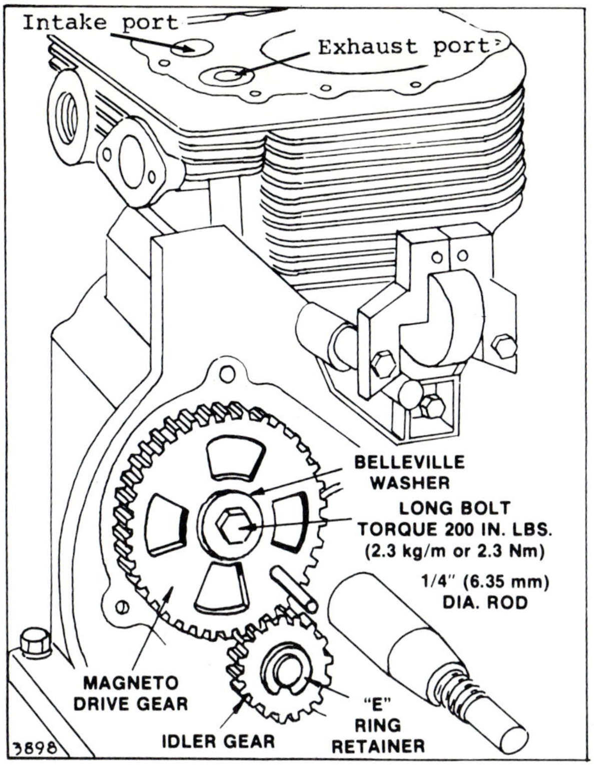

FIGURE 2-15. Cylinder head showing intake and exhaust ports (Courtesy of Briggs & Stratton Corporation).

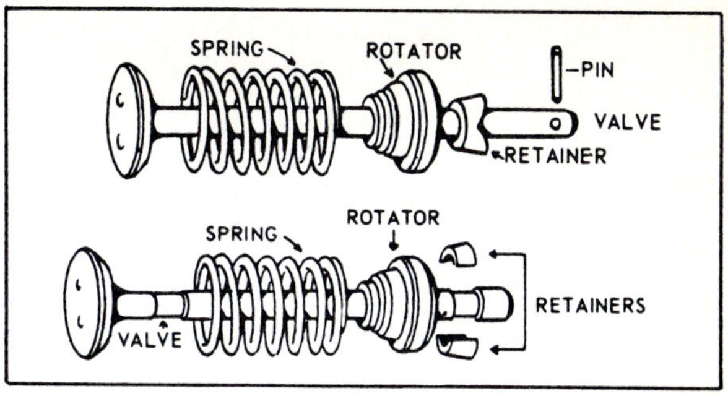

FIGURE 2-16. Valves, valve springs, and retainers (Courtesy of Briggs & Stratton Corporation).

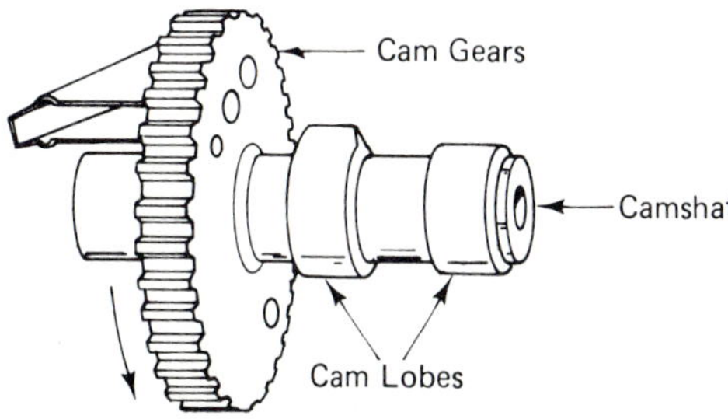

FIGURE 2-17. Cam gears, cam lobes, and camshafts (Courtesy of Clinton Engines Corporation).

mounted on the end of the crankshaft. The camshaft gear has twice as many teeth as the crankshaft gear, which means that the camshaft rotates at half the crankshaft speed.

OPERATING CYCLES

The term *cycle*, as applied to engines, means a complete series of events, from start to finish. More particularly, an engine cycle refers to those events responsible for: (1) delivering air and fuel inside the cylinder, (2) getting the air and fuel ready for burning, (3) burning the mixture, and (4) removing the exhaust gas. Gasoline engines (large and small) employ two basic operating cycles: four-stroke (Otto cycle) and two-stroke. In four-stroke cycle engines, four separate piston movements, or strokes, are required for the four events noted above—one stroke for each event. Two-stroke cycle engines combine the events into two piston strokes.

Piston Position Terms

A stroke refers to one complete up or down motion of the piston. Before examining four- and two-stroke operation, we need to define a few terms that are used to describe the position of the piston near either end of a stroke.

A piston at the top of its travel is said to be at TDC or Top Dead

Center. A piston at the bottom of its travel is said to be at BDC or Bottom Dead Center. "BTDC" means that the piston is on an upstroke, just before the top—or Before Top Dead Center. "ATDC," or After Top Dead Center, means that the piston has just moved past the top and is on its way down again. These terms will be used throughout the remainder of this book. (See Figure 2-18.)

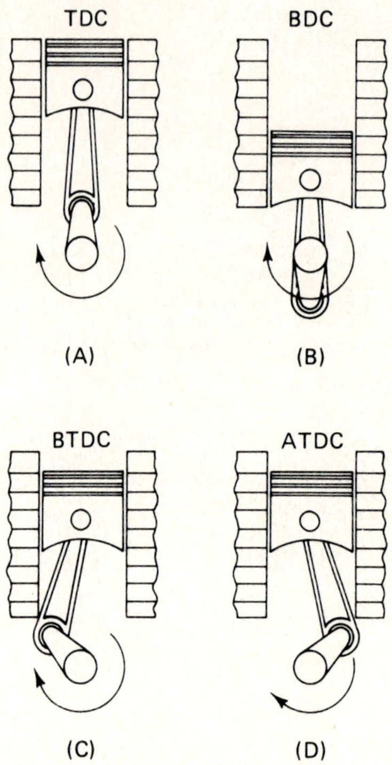

FIGURE 2-18. Piston position terms.

FOUR-STROKE CYCLES

As we noted above, four-stroke cycle engines require four piston strokes for each cycle. During each cycle, the piston makes two upstrokes and two downstrokes. Since two strokes equal one complete crankshaft revolution, or 360 deg of rotation, four strokes correspond to two crankshaft revolutions, or 720 deg of rotation.

The four strokes are called:

1. Intake
2. Compression
3. Power
4. Exhaust

Intake Stroke (180 deg of crankshaft rotation)

The intake stroke, which takes place as the piston moves from TDC to BDC, is responsible for delivering air and fuel inside the cylinder. (See Figure 2-19.) The process starts when the piston is near TDC. The intake valve is pushed open by a cam, providing an unobstructed path from the air cleaner and carburetor all the way to the cylinder.

Air and fuel are pushed along this path in response to pressure differences between the outside air and the air inside the cylinder. The pressure difference occurs as the piston moves down in the cylinder. Due to the increase in volume, the air inside the cylinder expands or "thins out" to create a partial vacuum. Air at normal atmospheric pressure on the outside of the engine pushes its way into the lower pressure region inside the cylinder. Fuel from the carburetor is also pushed into the air stream to create the required air/fuel mixture.

Compression Stroke (360 deg of crankshaft rotation)

When the piston nears the bottom of its travel in the intake stroke, the intake valve closes. Both the intake and the exhaust valves stay closed for the next part of the cycle, called the *compression stroke*. (See Figure 2-20.) As the name implies, the compression stroke is responsible for compressing the air and fuel mixture. This is necessary to obtain the proper energy release. By the time the piston reaches the top of its stroke, the volume over the piston (called the *combustion chamber*) is reduced as much as seven or eight times, packing the particles of air and fuel into a dense, hot mass that is ready to be ignited.

Power Stroke (540 deg of crankshaft rotation)

Combustion starts when an electric arc appears at the tip of the spark plug. The spark is timed to occur at the best moment for ignition, just before or just as the piston reaches TDC. When the fuel starts to burn, the air temperature and pressure inside the cylinder increase dramatically. The expanding gas pushes on all sides of the combustion chamber. However, both valves are still closed, and the only movable object is the piston. Responding to the force exerted on the piston head, the piston moves back down in the cylinder. The downward thrust is transferred to the connecting rod and the crankshaft. (See Figure 2-21.)

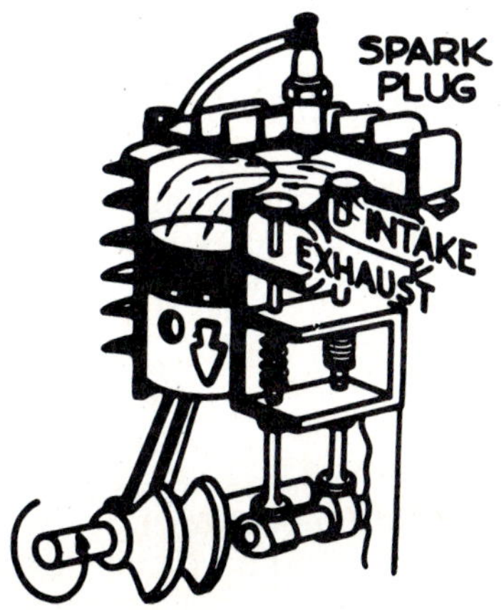

FIGURE 2-19. Intake stroke (Courtesy of Briggs & Stratton Corporation).

FIGURE 2-20. Compression stroke (Courtesy of Briggs & Stratton Corporation).

FIGURE 2-21. Power stroke (Courtesy of Briggs & Stratton Corporation).

Exhaust Stroke (720 deg of crankshaft rotation)

All that remains now is to clean the combustion by-products out of the combustion chamber. This is the job of the exhaust stroke. As the piston moves up from BDC to TDC, the exhaust valve opens. The exhaust gas contained in the cylinder is pushed out of the exhaust port, into the exhaust manifold and on to the muffler. Near the end of the exhaust stroke, the exhaust valve closes. At this point, the crankshaft has undergone 720 deg of rotation, and the next cycle is ready to begin. (See Figure 2-22.)

FIGURE 2-22. Exhaust stroke (Courtesy of Briggs & Stratton Corporation).

NOTES ON FOUR-STROKE OPERATION

To make the preceding explanation as simple as possible, we have neglected to mention a few, still-basic details that you'll need later on.

Cylinder Configurations

Although we said that the piston moves up and down in the cylinder, it should not be assumed that all cylinders are vertical. Some are horizontal. Also, many small engines have more than one cylinder. Some have two or four cylinders arranged in horizontally opposed pairs. Others are assembled in V-shaped configurations. However, no matter how they are arranged, the four-stroke principle applies to each cylinder. It takes two crankshaft revolutions, or 720 deg of rotation, to complete the operating cycle for each cylinder.

Flywheel Effect

Of the four strokes, only one, the power stroke, contributes power to the crankshaft. In order to keep the crankshaft turning smoothly during the other, "freeloading" strokes, a flywheel is provided. (See Figure 2-23.) The flywheel is a weighted wheel keyed or splined to the end of the crankshaft. Once put into motion (by the power stroke) momentum tends to keep the flywheel moving. This is due to the law of inertia, which refers to the

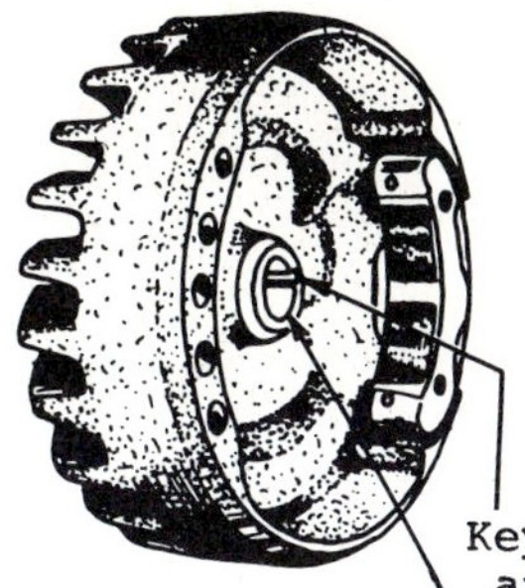

FIGURE 2-23. **Typical small-engine flywheel (Courtesy of Clinton Engines Corporation).**

tendency of objects to remain in motion and to resist any efforts to change their motion. The momentum or energy absorbed by the flywheel from the power stroke carries the crankshaft, piston, and piston rod through the other strokes. The energy left over is used to drive whatever the engine is attached to.

Valve Overlap

It may have been assumed from the previous discussion that the valves open and close exactly when the piston is at TDC or BDC. Actually, in many cases, the valves open a few degrees before the end of a previous stroke and close a few degrees after their own stroke ends. Such valve "overlap" is used to promote "scavenging," which will be explained later in this book.

TWO-STROKE OPERATION

Two-stroke engines must perform the same basic tasks as four-stroke engines. However, two-strokes must accomplish these tasks in one up and one down movement of the piston—in other words, in 360 deg of crankshaft rotation instead of 720 deg. The secret to getting everything done in two strokes is to use the area above *and* below the piston for moving air and fuel. To see how it works, let's trace the flow of air and fuel through one operating cycle.

Up Stroke

INTAKE

The carburetor is connected by a passageway to the crankcase in two-stroke engines. Flow through the passageway is controlled by a flapperlike reed valve. When the piston goes up in the cylinder, a low-pressure region is created in the crankcase (see previous discussion of four-stroke cycle intake strokes). Normal atmospheric pressure outside the carburetor pushes air and fuel through the reed valve into the crankcase.

During the next down stroke of the piston the pressure in the crankcase will increase. This will force the one-way reed valve shut, trapping the air/fuel mixture. The pressurized air/fuel mixture will remain in the crankcase until the later part of the down stroke, when it will then be transferred to the combustion chamber.

COMPRESSION

While air and fuel are being drawn into the region beneath the piston, a previous charge of air and fuel is compressed *above* the piston in the combustion chamber. The mixture is compressed in the usual manner as the piston moves from BDC to TDC. (See Figure 2-24.)

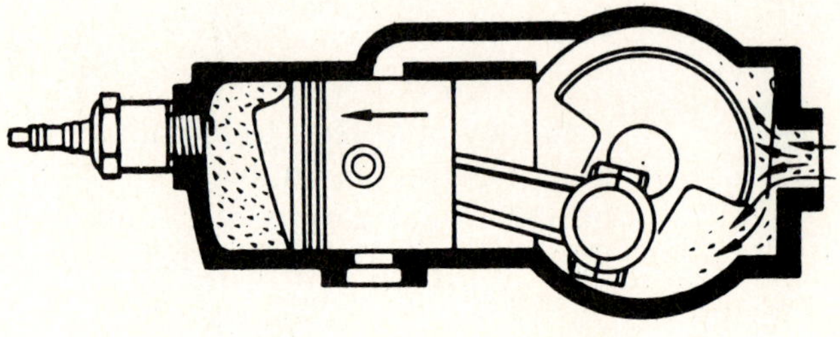

FIGURE 2-24. Two-cycle compression (Courtesy of Clinton Engines Corporation).

Down Stroke

POWER

When the piston reaches TDC, an arc occurs at the tip of a spark plug projecting into the combustion chamber. As in four-stroke engines, the air/fuel mixture is ignited and energy is released in the form of hot expanding gases. The gases push down on the piston, which transfers the force to the connecting rod and the crankshaft. However, the similarity with four-stroke engines ends near the lower part of the stroke. The retreating piston uncovers two holes or ports in the side of the cylinder. (See Figure 2-25.)

EXHAUST

The first hole is the *exhaust port*. When it is opened, the exhaust gases trapped in the cylinder start to escape into the exhaust manifold.

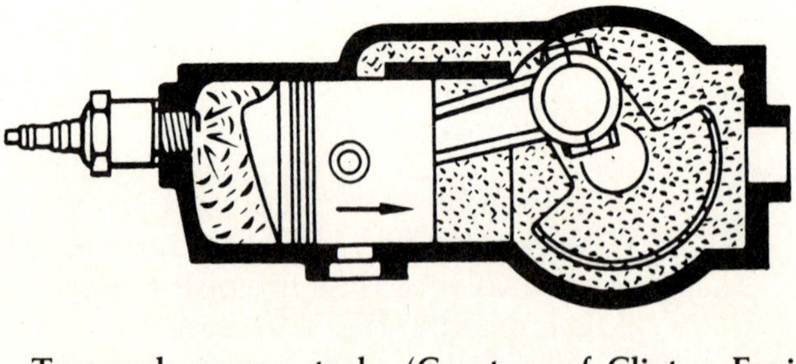

FIGURE 2-25. Two-cycle power stroke (Courtesy of Clinton Engines Corporation).

TRANSFER

A little later, a second hole, called the *intake* or *transfer port*, is uncovered. It is connected to a passage leading to the crankshaft. The pressurized air and fuel now escape into the combustion chamber. As the incoming mixture swirls through the cylinder, it forces out the last of the exhaust gas. After reaching BDC and starting back up in the cylinder, the piston covers the two ports, and the process starts again. (See Figure 2-26.)

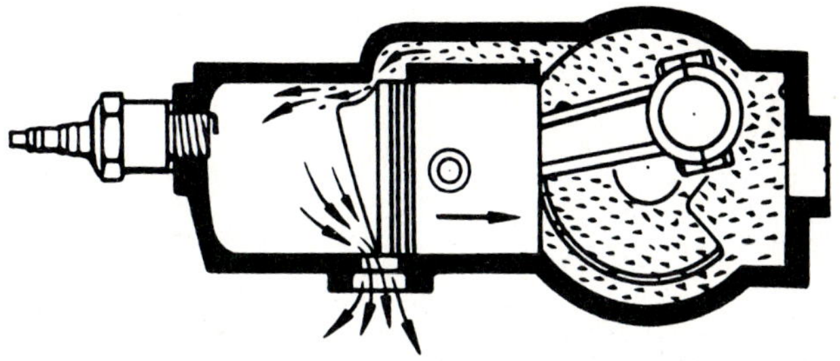

FIGURE 2-26. Two-cycle exhaust/transfer stroke (Courtesy of Clinton Engines Corporation).

3

PERFORMANCE MEASUREMENTS AND TERMINOLOGY

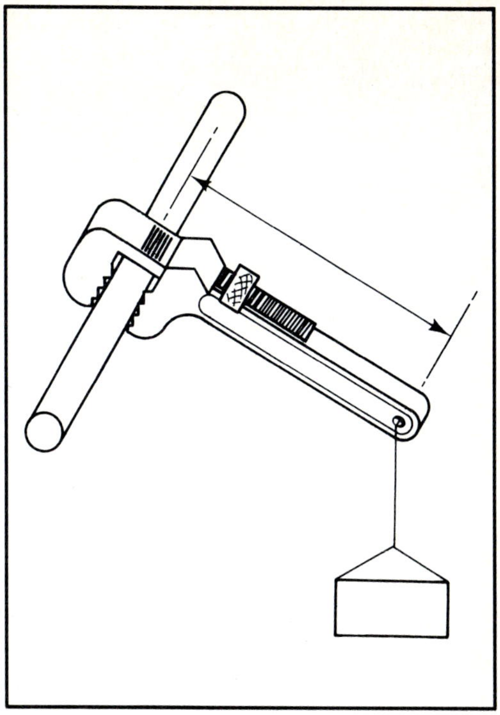

GENERAL

Whenever we talk about an engine, one of the first things we usually try to do is classify it: How big is the engine? How much power does it produce? What are the operating characteristics? This chapter explains the technically correct way to answer such questions. During the course of the explanation, such performance-related terms as work, energy, power, friction, etc. will be reviewed. By the end of the chapter, we will have constructed a framework of precise definitions that you can use later on when you compare the performance of small engines.

ENGINE SIZE/DISPLACEMENT AND TERMS

The size of an engine is often given in terms of its displacement, which, in multicylinder engines, means the combined working area of all its cylinders. Displacement is either expressed in inch units (e.g., cubic inches displacement, or CID) or metric units (CCs, liters, etc.) The following paragraphs show how displacement is determined.

Stroke

First, we need to know the stroke. It refers to the movement of a piston from TDC to BDC, or how far it travels from one end of the cylinder to the other. This figure is also given in inches or in metric units.

Crankshaft Throw

The stroke is directly related to the crankshaft throw—the offset of the crankshaft connecting rod journal with respect to the main bearing journal. For instance, if the offset is 2 in. (if the centerline of the main bearing journal is 2 in. from the centerline of the connection rod bearing journal), then the stroke will be twice that distance, or 4 in. If an engine has more than one cylinder, the throw and corresponding stroke are the same for every piston. (See Figure 3-1.)

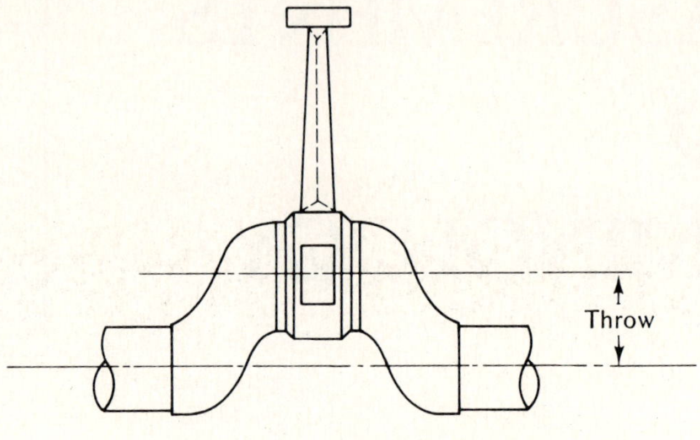

FIGURE 3-1. A crankshaft throw.

Bore

In addition to the stroke, we also need to know the bore. It is the diameter of the cylinder and is also the same for the cylinders in a multicylinder engine. Like the stroke, the bore is stated in inches or in metric units. The bore size also serves as a reference point in measuring for cylinder wear and in resizing a cylinder when an engine is being rebuilt. (See Figure 3-2.)

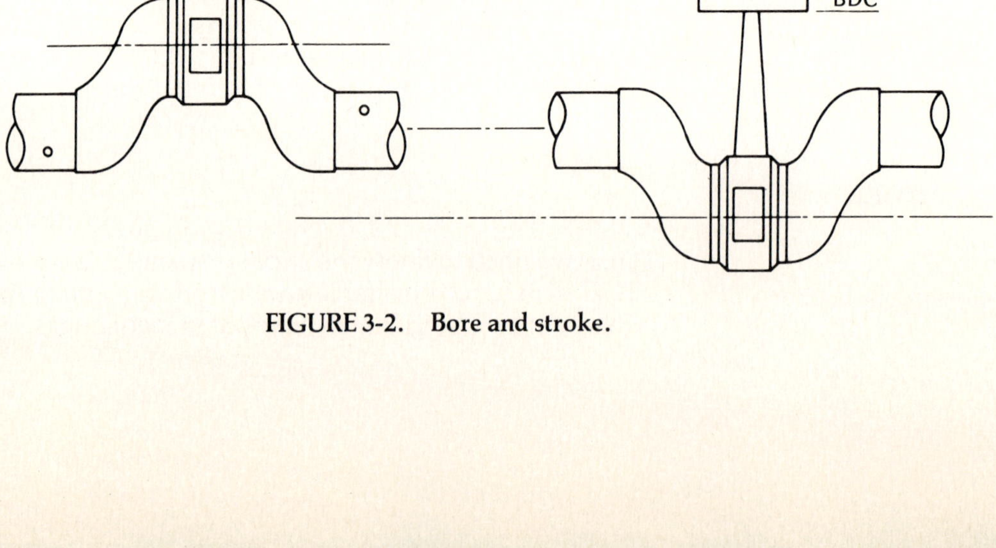

FIGURE 3-2. Bore and stroke.

Displacement

The bore and stroke are used to find the volume occupied by one piston as it moves from TDC to BDC. In other words, it is the volume of an imaginary cylinder whose height is the same as the stroke of a piston and whose diameter is the same as the bore of the cylinder. The formula for finding the displacement of a cylinder is the same as the formula for finding the volume of a cylinder.

$$\text{Volume} = \pi \times R^2 \times \text{stroke}$$

$$\text{or Volume} = 3.14 \times \tfrac{1}{2}\,\text{bore}^2 \times \text{stroke}$$

If the engine has more than one cylinder, the volume of the entire engine is obtained by multiplying the above figure times the number of cylinders in the engine. (See Figure 3-3.)

Volume

Occasionally the displacement of an engine is confused with the total volume. The total volume is the displacement *plus* the volume remaining in the combustion chamber when the piston is at TDC. This is sometimes called the clearance volume because it represents the clearance between the top of the piston and the head. (See Figure 3-4.)

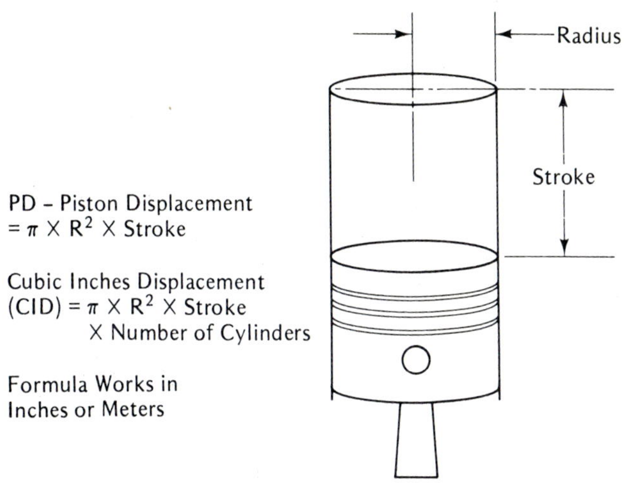

FIGURE 3-3. Formula for calculating cubic inch displacement (CID).

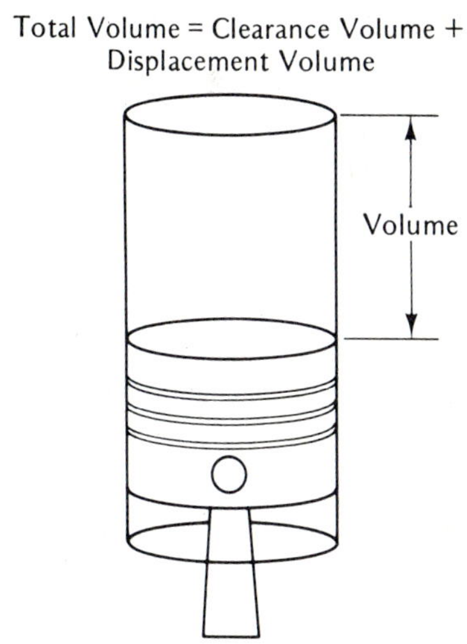

FIGURE 3-4. Cylinder volume.

Compression Ratio

As noted in Figure 3-5, the compression ratio is a comparison between two volumes. It is the ratio of relationship between the total volume in the cylinder and the clearance volume of the cylinder. The greater the ratio, the more air is squeezed in the compression stroke.

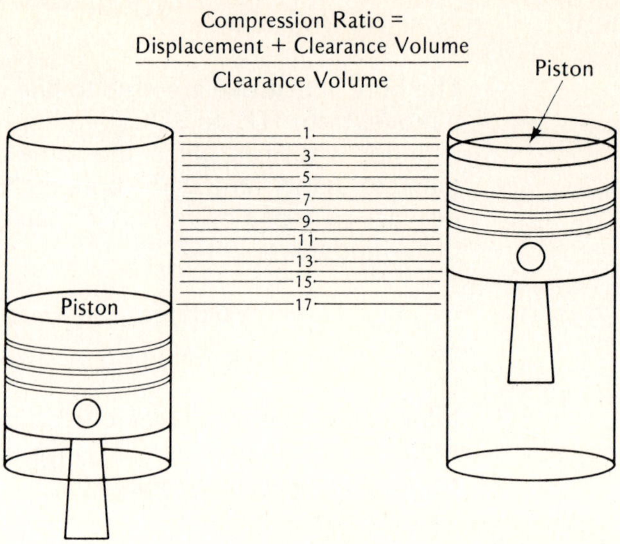

FIGURE 3-5. **Clearance volume and compression ratio.**

Volumetric Efficiency

This is another comparison between two volumes—in this case, between the volume of air that theoretically could enter the cylinder and the volume of air that actually does enter the cylinder. It is a measure of how efficiently the engine breathes.

The ratio is based on a standard air temperature of 60°F at sea level pressure (14.7 pounds per square inch). The volumetric efficiency of most well-designed small engines is on the order of 80 percent at moderate speeds. At higher speeds the volumetric efficiency might drop to 50 percent.

BASIC PHYSICAL CONCEPTS

Before we see how power is measured, we need to briefly examine the basic concepts of work, force, energy, power, etc.

Force

Force is an effort exerted against a resistance. When you push against a wall, you exert force.

Work

Work is a combination of force and movement. If you push against a wall and the wall doesn't move, you are not doing any work, no matter how hard you push. However, if the wall moves, then you are doing work. The typical unit of measuring work is the foot-pound. If a ten-pound object is moved 10 feet, 100 foot-pounds of work have taken place.

Energy

Energy is the ability or potential to do work. Fuel possesses chemical energy, which is converted by the engine into mechanical energy. Mechanical energy is transmitted from the piston to the crankshaft and on to whatever the engine is operating. Although energy can neither be created nor destroyed (except in thermonuclear devices), it can be degraded into lower and lower forms until it can no longer be put to practical use. Heat energy radiated from the exhaust is an example of low-grade, wasted energy.

Torque

Torque is force applied in a circular or twisting manner. When you turn or twist on a screwdriver handle, you are exerting torque. The same idea applies when the crankshaft is rotated by the piston and connecting rods. Torque is measured in pounds-feet (not foot-pounds). Figure 3-6 illustrates torque and how it is measured.

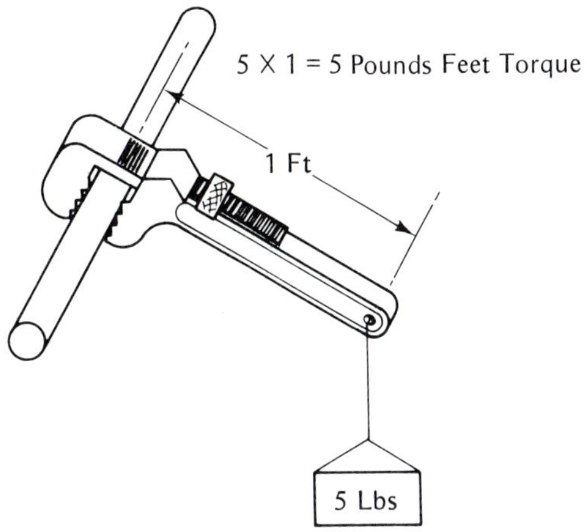

FIGURE 3-6. Torque means turning or twisting force.

Inertia

If you have ever pushed a car or heavy wagon by hand, you probably noticed that it was more difficult to get the vehicle moving than it was to keep it moving. You have probably also noted that once the vehicle is moving, it displays a stubborn tendency to go in the direction it is headed. It has to be forced to go around corners, forced to slow down, forced to speed up. This is inertia, the resistance of an object to changes in speed or direction. An engine flywheel makes use of inertia by resisting changes in speed to dampen out uneven firing impulses and to help balance the engine. (See Figure 3-7.)

FRICTION

Friction is either friendly or unfriendly, depending on where it is found, and on what kind of friction it is.

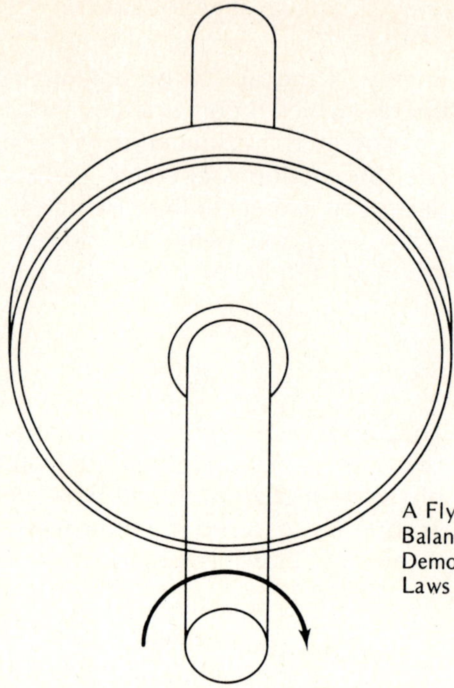

FIGURE 3-7. The laws of inertia control the operation of a flywheel.

Dry Friction

This is just what the name says, resistance to sliding movement between two dry surfaces. In a car, the friction between the brake shoe and the drum, or between the brake pad and rotor, is an example of dry friction. Without this kind of friction, the car could not stop. Without dry friction between the tires and the road, a car could not move forward or be kept on the road. In these cases, dry friction is a friendly phenomenon. However, dry friction isn't so friendly when it takes place between two moving or rotating parts inside the engine—for instance, between a journal and a bearing surface.

Greasy Friction

As much as possible, dry friction is eliminated between rotating or sliding engine parts. You might think that dry friction could be eliminated by supplying a layer of grease or heavy lubricant between the bodies. Up to a point this is true. However, at high speeds, thick grease separates and lets the engine parts come into contact. As a result, friction and wear between the parts is not reduced appreciably.

Viscous Friction

This is the friction or resistance to movement that occurs between the molecules of a liquid. If the proper lubrication is introduced between two high-speed sliding or rotating bodies, the bodies themselves will not touch and friction will be limited to resistance to motion between layers of the lubricant. That is basically how oil protects the engine parts. (See Figure 3-8.)

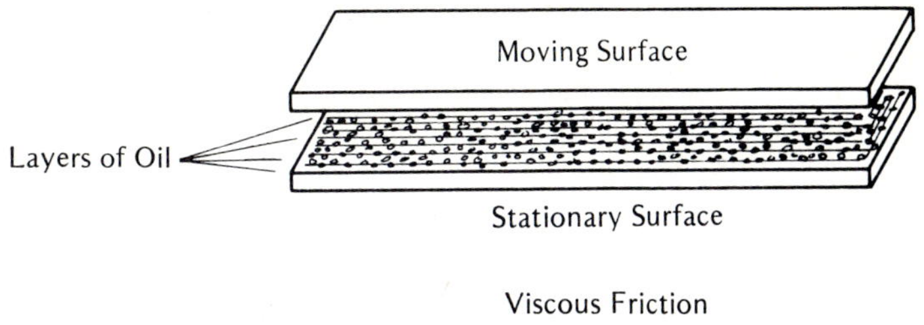

FIGURE 3-8. Viscous friction between molecules of a liquid.

POWER

Everybody has an idea of what power is. We say a person is powerful (physically, mentally, emotionally, etc.). We describe a machine as being powerful. In each case, we are describing the ability of someone or something to perform in certain ways. The technical definition of power is much the same, except that we add one ingredient that might not have been present in our informal definition—time. In a technical sense, power is the ability to do a certain amount of work in a certain period of time. There are a number of ways to measure power and various units used to express how much power a device can produce. The following paragraphs describe some ways used to measure the power produced by internal combustion engines.

Horsepower

This is the most common way to express the power produced by an engine. The term originated when draft horses were still in common use. Engineers and scientists rather arbitrarily decided that one large horse could lift 33,000 pounds one foot in one minute. So an engine that could do the same amount of work in the same time was said to produce one horsepower.

The problem is how do you measure horsepower? How do you find out how much work an engine can do in a certain period of time? It is not practical to match an engine with a team of horses. Some other method must be used. The next paragraphs note some of the methods that have been used.

SAE Horsepower

One of the earliest ways to describe the horsepower of automobile engines was developed by an organization called the Society of Automotive Engineers (SAE). They didn't actually measure how much work an engine does over a period of time, but instead used this formula to indirectly compare the output of engines:

$$\text{SAE hp} = \frac{\text{bore diameter}^2 \times \text{number of cylinders}}{2.5}$$

This formula is simple; anyone can use it to find how much power an engine can produce. Unfortunately, though, the formula doesn't take into account such factors as compression ratio, engine speed, length of stroke, etc. As a result, SAE horsepower is rarely used except for legal or licensing purposes.

Brake Horsepower (BHP)

Brake horsepower is a direct measurement of the usable power produced by an engine. It is the most common way to determine the output of an engine.

The term *brake*, as related to horsepower, originated with a device called the Prony Brake. It was a forerunner of the modern dynamometer and consisted of a variable braking device connected by a lever to a weighing scale. The arm resting on the scale applied resistance to a wheel turning inside the brake. The engine being tested was connected to the wheel. The ability of the engine to turn the wheel at certain revolutions per minute (RPMs) and under certain loads was measured and the resulting figures inserted into a formula to determine its brake horsepower. Figure 3-9 pictures a Prony Brake and gives the formula for determining brake horsepower.

Today, brake horsepower can be checked at the end of the crankshaft by using an engine dynamometer. Some modern dynamometers are absorption units and work on the principle of the Prony Brake. Other dynamometers are actually electrical generators connected to the engine. The brake horsepower of the engine is determined by measuring the amount of

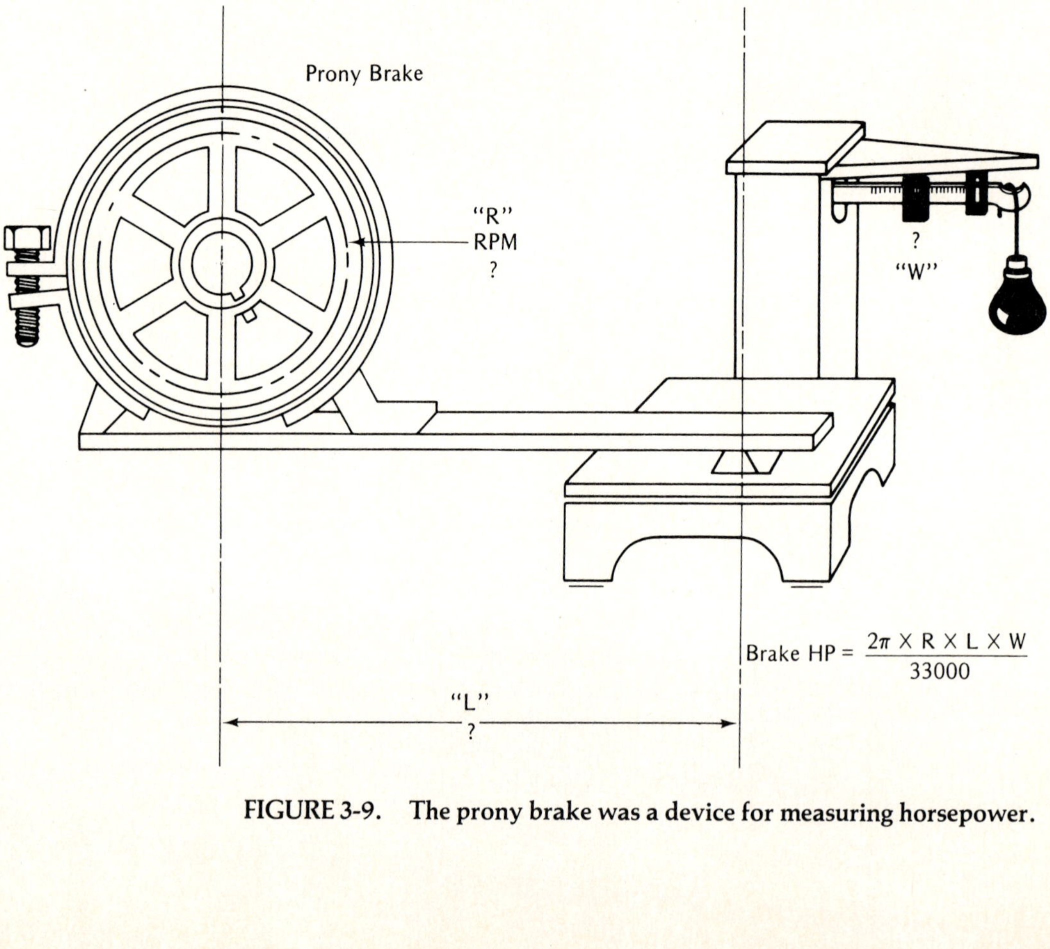

FIGURE 3-9. The prony brake was a device for measuring horsepower.

electrical power produced (746 watts = 1 hp) and converting that figure to brake horsepower.

SPECIALIZED HORSEPOWER RATINGS

In addition to these common ways of measuring and stating horsepower, engineers use special measuring techniques to describe and compare various aspects of engine operation.

Indicated Horsepower

Indicated horsepower is a relative measurement of the power produced by burning fuel in an engine's cylinders. First, the mean effective pressure produced during the four strokes of operation is determined by electronic testing devices in a laboratory. Then that reading is inserted into the following formula:

$$\text{Ihp} = \frac{P \times L \times A \times N \times K}{33{,}000}$$

where P = mean effective pressure in pounds per square inch (psi)

L = length of the stroke in feet

A = cylinder area in square inches

N = power strokes per minute $\frac{\text{RPM}}{2}$

K = number of cylinders

Frictional Horsepower

This is a measurement of the power lost due to friction—in other words, to the resistance of engine parts in sliding contact. The test is performed by driving an engine by an electrical motor and measuring the resistance offered. Tests have shown that friction increases with speed and that as much as 70 to 80 percent of friction losses occur in the cylinder area.

TEMPERATURE/HEAT

Temperature and heat are important concepts in engine operation. Temperature measurements are constantly being made by various devices, and automatic adjustments are made on the more expensive small engines. Heat relates to the basic operation of the engine itself, since the engine is a device for converting chemical energy into heat energy, which is then converted into mechanical energy.

Atomic Structure of Matter

Before we talk about the related concepts of heat and temperature, it is important to remember that all substances—you, this book, all materials in the universe—are made up of tiny particles called atoms. Atoms, in turn, are made up of even smaller particles called electrons, protons, and neutrons. Heat and temperature are measurements of the activity within the atoms of matter.

Temperature

Atoms and their component parts are always in motion. Temperature is a measure of that motion. High temperature means more atomic activity; lower temperatures mean less activity.

Heat

Heat is a measurement not only of how much atoms vibrate, but of how many atoms are in vibration. In other words, heat is a measurement of mass as well as activity. Look at it this way: A large body of water at 33°F contains more heat than a small kettle of boiling water at 212°F. There is more total activity (atomic) in the large body of water than in the small kettle.

Temperature Measurement Scales

There are two common scales for measuring temperature: fahrenheit and celsius. Both are based on the freezing and the boiling points of water at sea level atmospheric pressure (14.7 pounds per square inch or psi). In the United States, the fahrenheit scale of 32 (freezing) to 212 degrees (boiling) is most commonly used. However, the trend is toward the metric or celsius method, which sets the freezing point of water at 0 degrees and the boiling point at 100 degrees.

Heat Measurement Scales

Since heat is a measurement of quantity as well as activity, it must take into account mass as well as temperature. There are two common ways to measure heat: the BTU (British Thermal Unit) and the calorie. The BTU is the heat required to raise the temperature of one pound of water one degree fahrenheit. (See Figure 3-10.) A calorie is the heat required to raise the temperature of one gram of water one degree celsius (actually from 3.5 to 4.5°C). This means that if you eat a doughnut containing 100 calories, you have consumed enough food energy to raise one gram of water 100 degrees—from freezing to boiling.

TEMPERATURE MEASUREMENT DEVICES

The most common temperature measurement device is a bulb thermometer like the one shown in Figure 3-11. It consists of a glass bulb containing liquid mercury or colored water and a graduated column connected to the bulb. The bulb is placed near the substance whose temperature is being measured (outdoors in the air, under the tongue, in water, etc.). Atomic ac-

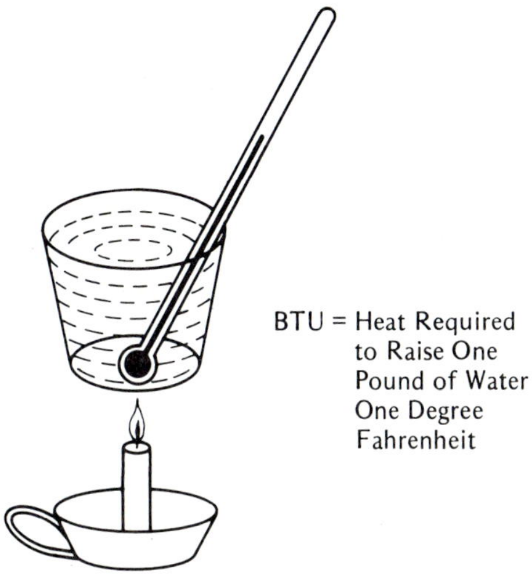

FIGURE 3-10. BTUs are units of heat measurement.

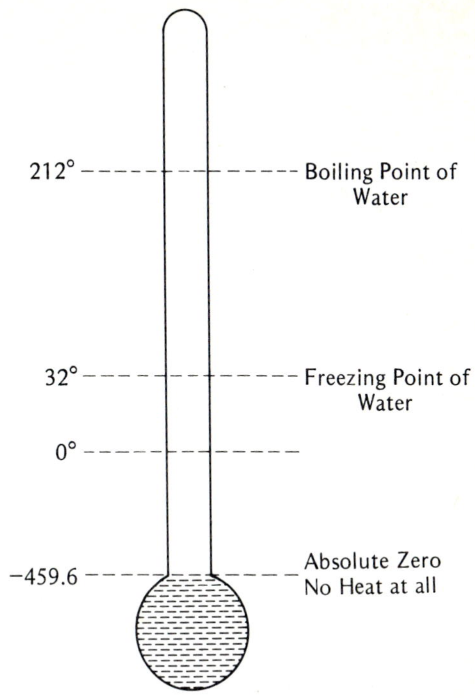

FIGURE 3-11. A glass bulb thermometer.

tivity in the substance being checked is transferred to the liquid in the bulb, causing the liquid to expand or contract. As the liquid changes volume, it goes up or down in the column. The height of the liquid is compared to the graduated scale and read as temperature.

Although thermometers used on engines are hardly ever of the glass bulb type, they still respond to atomic activity to indicate temperatures.

Thermocouple

This kind of temperature-sensing device consists of two dissimilar wires welded together at one end and connected to a sensitive meter at the opposite end. (See Figure 3-12.) Heat applied to the ends joined together creates atomic activity that the meter registers as electrical current flow. By properly calibrating the meter, electrical readings can be converted to temperature readings. Heat gauges of this type are more often found on liquid-cooled engines and are not used extensively on small engines.

Bimetal Thermometers

The bimetal coil or spring is another kind of temperature-sensing device that makes use of dissimilar metals joined together. In this case, strips of dissimilar metals are bonded, sandwichlike, into layers. When the bimetal thermometer is exposed to heat, the two metals absorb atomic energy and expand. However, because their atomic structures are different, the two metals expand at different rates. As a result, the strip tends to curl, or, if

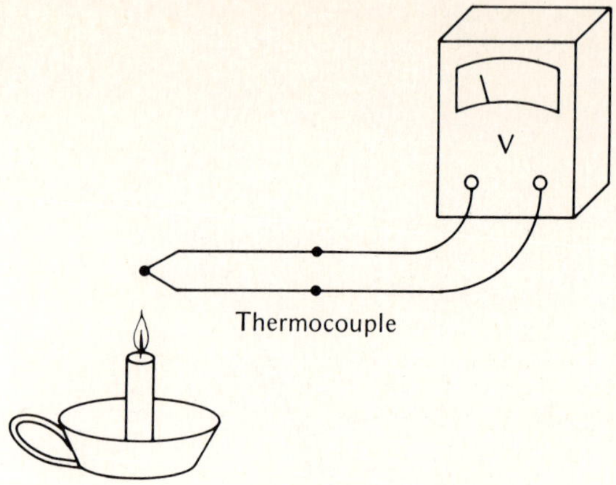

FIGURE 3-12. Diagram of a thermocouple.

fashioned into a coil shape, to wind up or unwind. (See Figure 3-13.) Bimetal-type thermometers are usually used to signal temperature changes to some type of electrical or mechanical device. The bimetal strip is attached to a set of electrical contacts or a small lever or an indicator pointer calibrated in degrees.

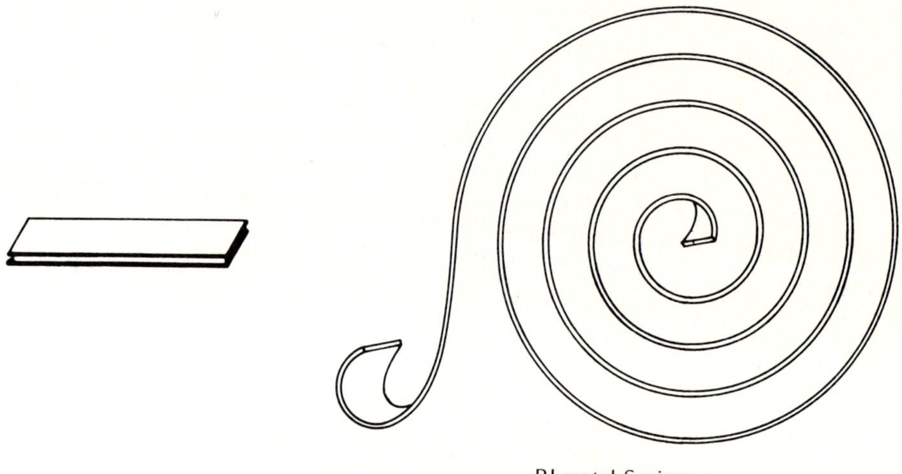

FIGURE 3-13. Bimetal springs react to changes in temperature.

EXPANSION

As noted in the previous paragraphs, when materials absorb atomic activity from other, more active sources, expansion occurs. The fractional amount of metal expansion for a one-degree rise in temperature is called its *coefficient of expansion*. Expansion rates of all materials used are taken into account by engineers when they design an engine. Otherwise the expanding parts could seize (stick together) and destroy the engine.

4

SMALL ENGINE CONSTRUCTION

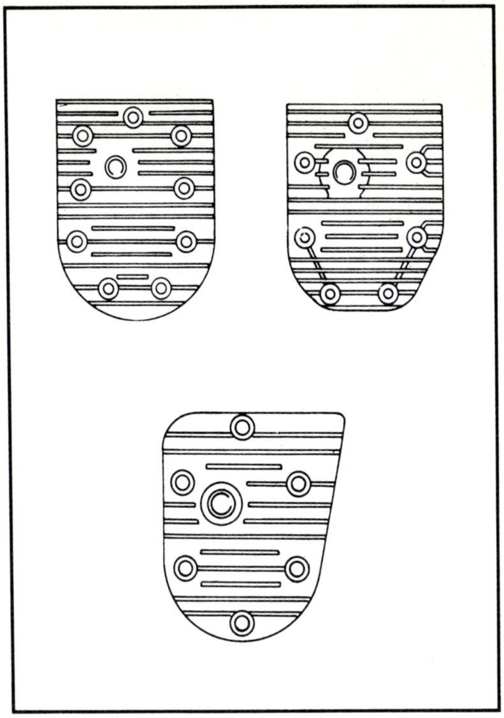

GENERAL

This chapter examines the design and construction of the major components used in small gasoline-powered engines. Some of the components introduced in this chapter will be covered again in later, service-oriented chapters.

TYPICAL ENGINES

As noted in Figure 4-1, there are various types of small engine configurations. However, a close examination of these engines will show that most are similar in basic design and construction. Therefore, this chapter will concentrate on the "typical" small engine, e.g., an air-cooled, single-cylinder unit employing the 4-stroke operating cycle. Understanding the construction of such a typical engine will give you a logical point of departure for understanding other types of small engines.

BLOCK

The block is the framework of any engine (typical or otherwise). Its shape defines the location and operation of all other engine components.

Blocks are usually (but not always) cast in one piece by pouring molten iron or aluminum into molds made out of sand. Blocks from cast iron are stronger, but weigh more. Aluminum blocks, besides being lighter, are easier to machine and they dissipate heat faster. However, aluminum,

FIGURE 4-1. Various types of small engines (Courtesy of Onan).

because of its wear rate, provides a poor cylinder surface. Tube-shaped cylinder liners must be used. Liners, which are made from steel, are pressed into the cylinder bore before the block is completely finished. That way, the liners can be sized and honed (polished) at the same time the block is machined to final tolerances.

Both iron and aluminum blocks are cast with integral cooling fins. The fins increase the amount of block surface area exposed to the air. Heat from the cylinders is dissipated into the fins, then radiated from the fins into the passing air stream.

The lower area of the block is called the *crankcase* because it encases or provides an enclosure for the crankshaft. (See Figure 4-2.) The crankcase may also house the camshaft and provide a sump or reservoir for the lubricating oil. The exact use of the crankcase depends on the particular type of small engine. For instance, a two-cycle engine would *not* use the crankcase as an oil reservoir.

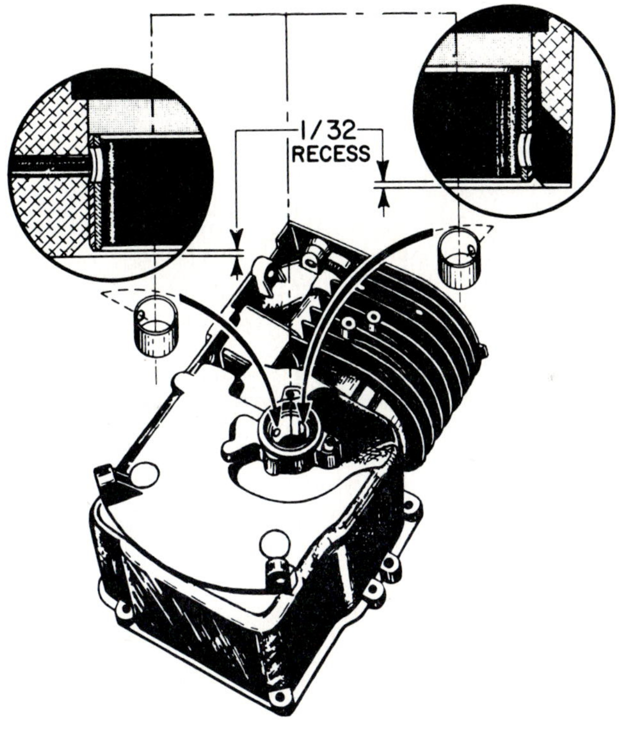

FIGURE 4-2. The crankcase is the lower part of the block (Courtesy of Clinton Engines Corporation).

CRANKSHAFT

If the block defines the framework of an engine, then the crankshaft is the focus of the engine's operation. That is because all mechanical activities come together at the crankshaft to produce turning or twisting force. (See Figure 4-3.) Some crankshafts are forged or stamped from a hot steel billet (single piece of metal) and then machined into final dimension. Others are cast by pouring molten metal into a mold. (See Figure 4-4.)

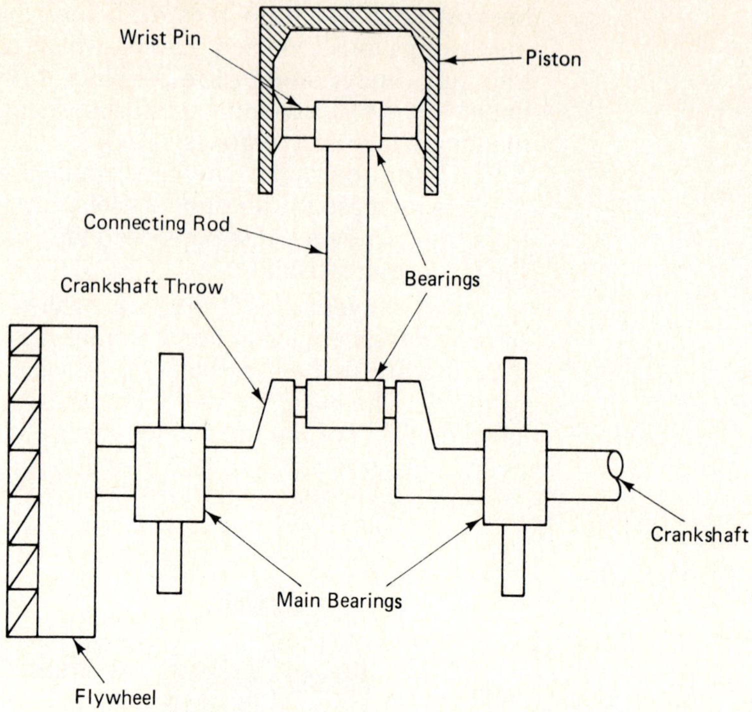

FIGURE 4-3. **The crankshaft and other major moving components in a small engine.**

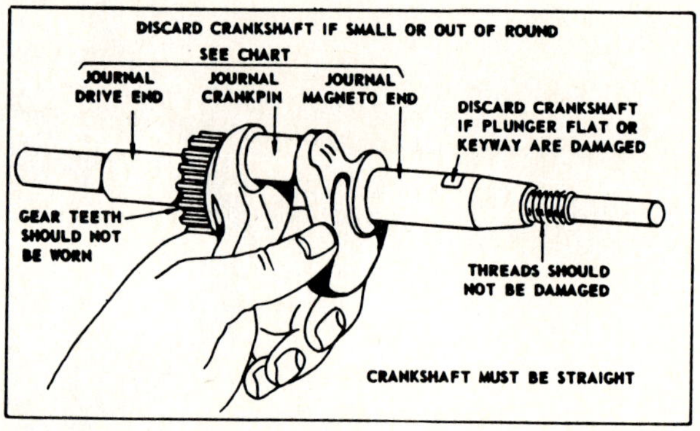

FIGURE 4-4. **Check points on a typical one-cylinder crankshaft (Courtesy of Briggs & Stratton Corporation).**

MAIN BEARINGS

Because all operations that require rotational power are connected in some way to the crankshaft, it must be well supported. The main bearings hold the crankshaft in position while allowing it to rotate with a minimum of friction.

Main bearings are usually pressed into matching holes formed on opposite sides of the crankcase. Reinforcing ridges, called *struts*, are cast into the block around the bearing holes.

Two kinds of bearings are generally used: bushing and roller type. Bushings are somewhat doughnut-shaped and are constructed from soft metal (brass or other alloys). The ends of the crankshaft rotate within the bushings in a sliding manner. Roller bearings reduce friction by providing a moving contact between the crankshaft and the bearing surface. Roller bearings, which are more expensive, are likely to be found in high-speed engines. Bushing-type bearings are adequate for most lower-speed applications.

SEALS

Since both ends of the crankshaft extend through the block, seals as well as bearings must be used. Seals are most often constructed from soft rubber. By fitting tightly between the crankshaft and the block, seals prevent oil from leaking around the ends of the crankshaft. Seals also prevent oil and dirt from entering the crankcase.

FLYWHEEL

A flywheel is usually mounted on one end of the crankshaft. (See Figure 4-5.) The end of the crankshaft is tapered so that a locking nut tightened from the end will pull the flywheel onto the taper for a tight fit. The end of the crankshaft also has a key which projects above the crankshaft surface. When engaged with a slot in the bore of the flywheel, the key prevents the flywheel from rotating independently of the crankshaft.

Flywheels serve two main purposes in small air-cooled engines. First, the momentum built up by the rotating flywheel tends to keep the engine moving between power strokes. As a result, uneven firing impulses have less effect on speed and performance. Normal engine roughness is also reduced. For this reason, flywheels, whether made of iron or aluminum, are relatively heavy.

Flywheels are also used in engine cooling. Fins cast in the upper surface of the flywheel force air to move over the engine. Often, a sheet metal shroud is placed around the flywheel to help direct the flow of cooling air (as well as to keep the operator's fingers from becoming destructively involved with the flywheel).

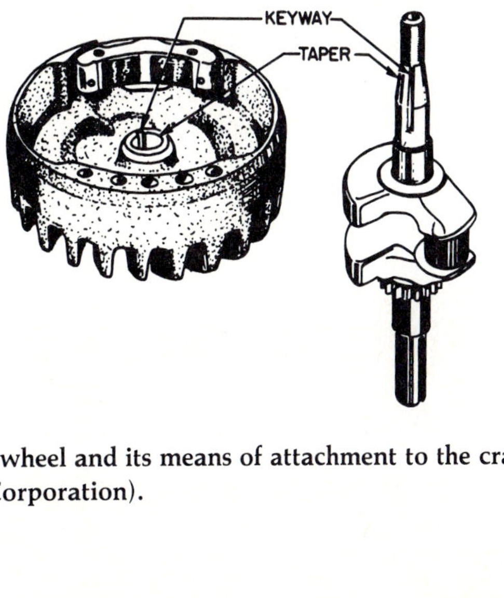

FIGURE 4-5. A flywheel and its means of attachment to the crankshaft (Courtesy of Clinton Engines Corporation).

CONNECTING ROD

The piston reciprocates back and forth in an up and down manner. The crankshaft rotates around its axis. The connecting rod joins the two together, changing the reciprocating motion of one into the rotation of the other without hindering the operation of either. [See Figure 4-6(a).]

In order to perform this task, the connecting rod must pivot at both ends. The upper pivot point is a precisely machined hole fitted with a bushing or bearing. The piston pin slides into this bearing surface.

The lower pivot point is sometimes called the "big" end of the rod. It provides a movable connection with the crankshaft offset ("throw"). The bearing used on the lower end of the rod may be formed integrally with the rod material, or it may be a separate bushing or roller bearing. If not part of the rod itself, the bearing is held in place by a removable cap. The cap, which is attached by bolts, allows the rod to be easily removed or installed. [See Figure 4-6(b).]

Rods usually have an "I" beam cross section for added strength. Some rods are forged. Others are cast by pouring molten steel or aluminum into molds.

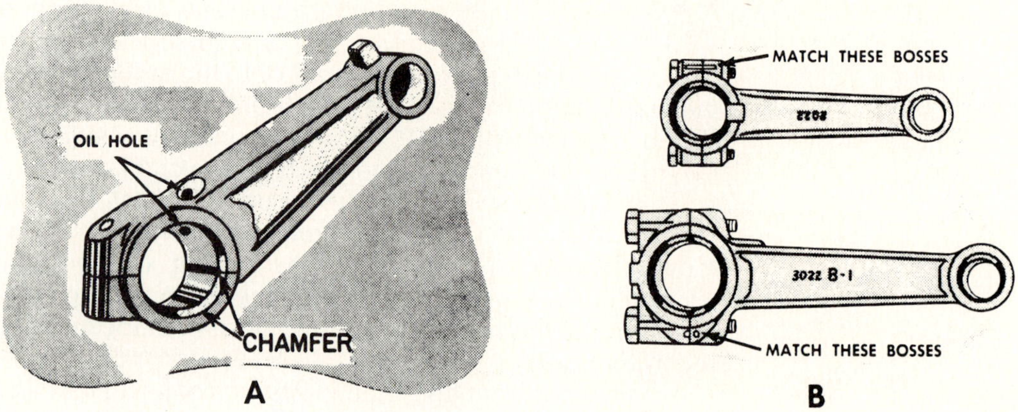

FIGURE 4-6. (A) A typical connecting rod; (B) connecting rods showing assembly marks (Courtesy of Clinton Engines Corporation).

PISTONS AND RINGS

Pistons in most modern small engines are cast from aluminum, then machined to final tolerances. The upper piston area is called the *head*. The lower part is called the *skirt*. (See Figure 4-7.)

Pistons must perform two somewhat contradictory operations. On the one hand, a piston must provide an effective seal with the cylinder walls so that expanding gases from the combustion chamber do not escape (and so that air is properly compressed during the compression stroke). On the other hand, a piston must move with a minimum of friction against the cylinder. Otherwise, excessive wear and a loss of power would result.

The problem is resolved by machining the piston a few thousands of an inch smaller than the cylinder, and inserting circular rings into grooves near the piston head. The rings slide easily to minimize friction and, at the same time, press against the cylinder walls to create a seal.

Three rings are normally used. The top two are compression rings, designed to seal the combustion chamber. The bottom ring (called the *oil control ring*) regulates the oil splashed on the cylinder walls by the lubrication process. Rings are constructed from iron or steel. Some rings are coated to help reduce wear and friction. (See Figure 4-8.)

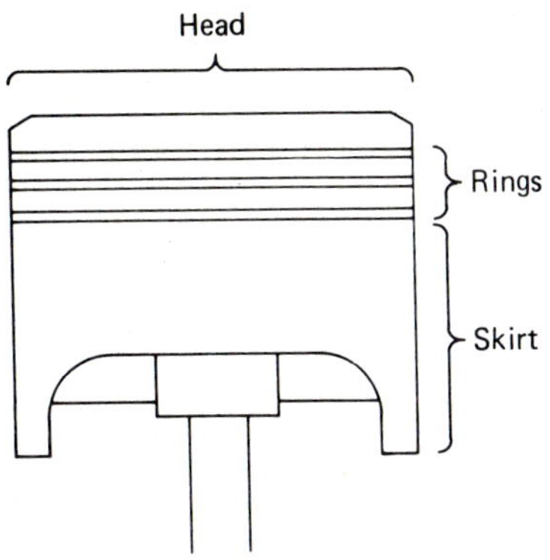

FIGURE 4-7. Piston nomenclature.

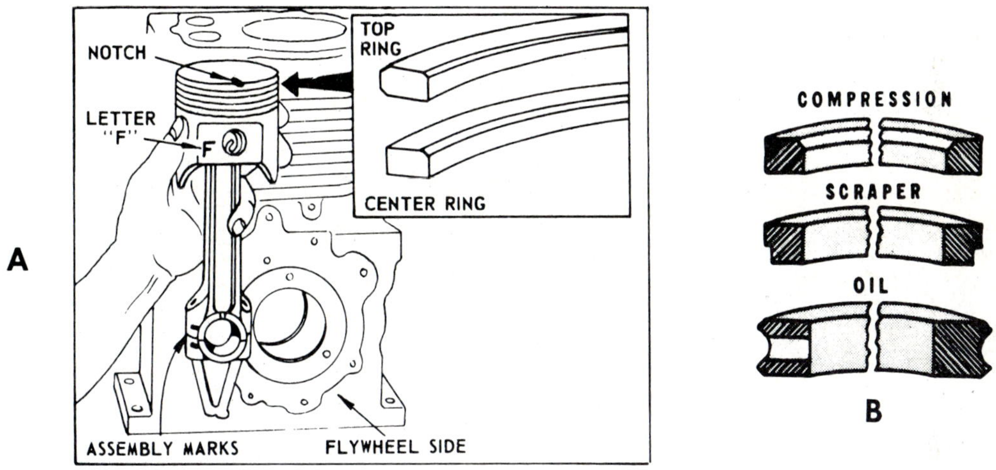

FIGURE 4-8. (A) Piston, rod, and rings; (B) cross-section of small engine piston rings (Courtesy of Briggs & Stratton Corporation and Clinton Engines Corporation).

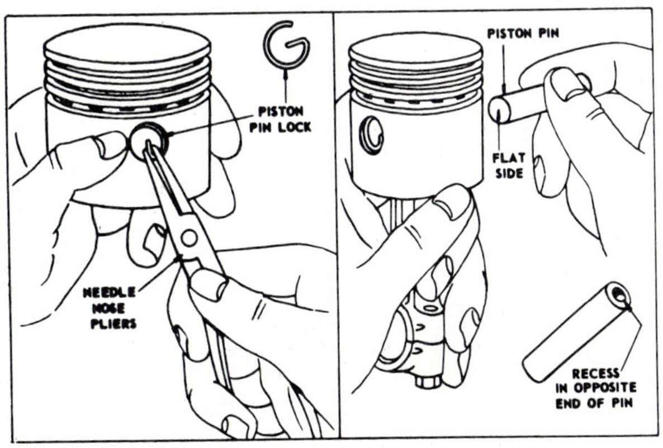

FIGURE 4-9. Pistons, piston pins, and piston retainer snap rings (Courtesy of Briggs & Stratton Corporation).

The lower part of the piston, or skirt, stabilizes the piston as it moves up and down in the cylinder. The skirt also serves as a location for the piston pin. The pin (sometimes called the *wrist pin*) fits in reinforced holes (bosses) drilled through both sides of the skirt parallel to the piston head.

The piston pin is inserted through the upper end of the connecting rod. The pin is made of specially hardened steel and polished to a high degree so that the rod can rotate freely. In most cases, the pin itself also rotates inside the piston. It is held in lateral position by snap rings that fit into grooves machined in the piston bosses. (See Figure 4-9.)

CYLINDER

We have already noted that the cylinder is a hole in the block where the piston moves up and down. We have also noted that aluminum blocks are likely to employ a steel cylinder liner to improve wear. What we haven't pointed out is that the cylinder walls are often honed with special tools to give a criss-cross hatch pattern like that shown in Figure 4-10. This improves wear and enhances piston performance. (See Figure 4-11.)

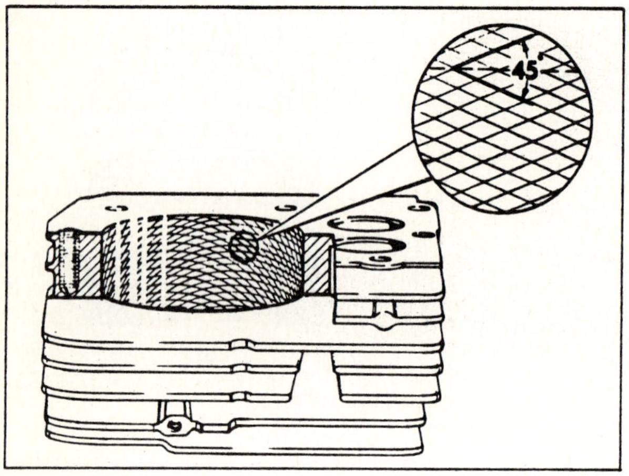

FIGURE 4-10. Crosshatch cylinder finish (Courtesy of Briggs & Stratton Corporation).

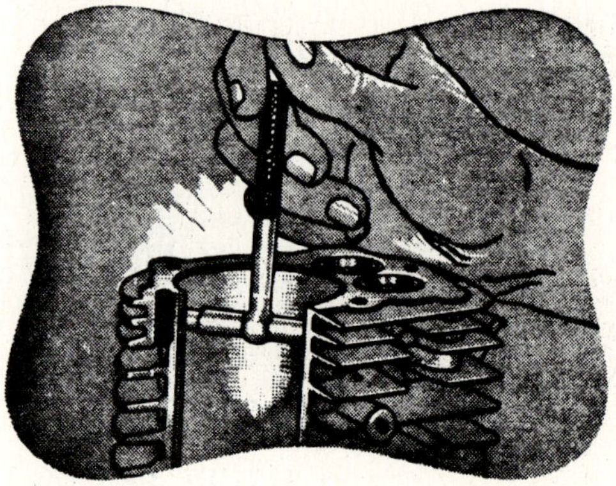

FIGURE 4-11. Checking the size of a cylinder (Courtesy of Clinton Engines Corporation).

VALVE TRAIN

The valve train is responsible for getting the air fuel mixture into the cylinder at just the right time, then providing an opening for removal of the exhaust gas when the combustion process is completed. (See Figure 4-12.) Chapter 2 introduced the operation of the valve train. This chapter will concentrate on the design and construction of the valve train components.

Valves

Valves are made from high-quality steel, ground to precise specifications. With reference to Figure 4-13, the top of the valve is called the *head*. The edge adjacent to the head is called the *margin*. Next comes the *face*. It is the

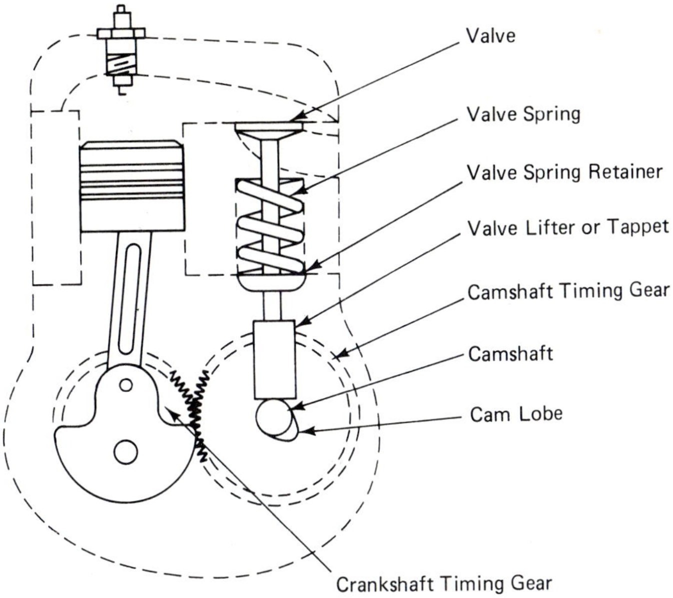

FIGURE 4-12. **Valve train schematic.**

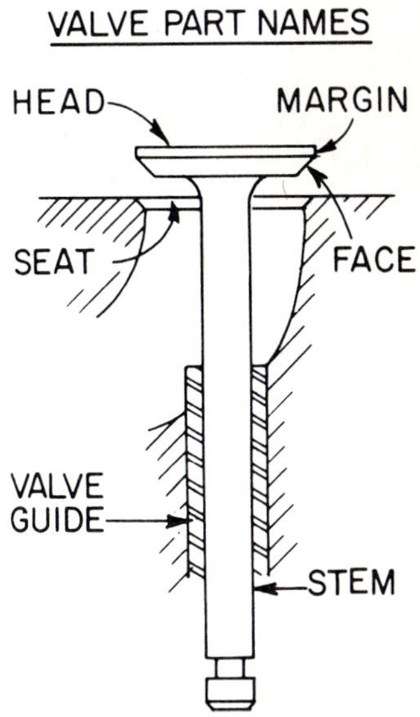

FIGURE 4-13. **Parts of a valve plus valve stem and seat (Courtesy of Clinton Engines Corporation).**

part of the valve that contacts the valve seat—where the port opens into the cylinder. Finally, the long part of the valve is referred to as the *stem*.

Since the valve is opened by the action of the camshaft and closed by a valve spring, the valve must be attached to the spring in some manner. An annular (circular) groove or small pin hole near the end of the valve stem is used in this connection. A pin inserted through the hole, or a "keeper" located in the groove prevents a spring retainer (a washerlike device) from slipping off the end of the stem.

Valve Spring

The spring retainer is what actually connects the spring to the valve. The spring pushes on the retainer, which in turn pulls on the valve. (See Figures 4-14 and 4-15.)

Valve Guide

The valve guide is a cylindrical passage in which the valve stem is located. The guide may simply be a hole drilled in the material surrounding the valve port, or the guide may be a piece of tubing pressed into the opening. Tubing-type valve guides are usually constructed from iron or steel. (See Figure 4-13.)

Camshaft

The camshaft is provided with one eccentric (out-of-round) surface or lobe for each valve—intake and exhaust. As the camshaft rotates, the lobes force

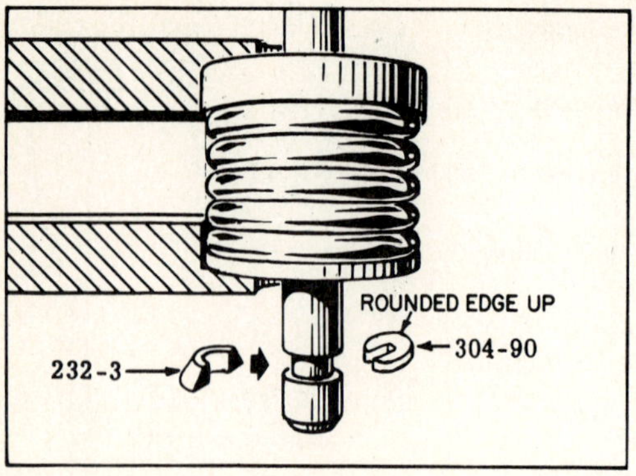

FIGURE 4-14. Valve spring and retainer (Courtesy of Clinton Engines Corporation).

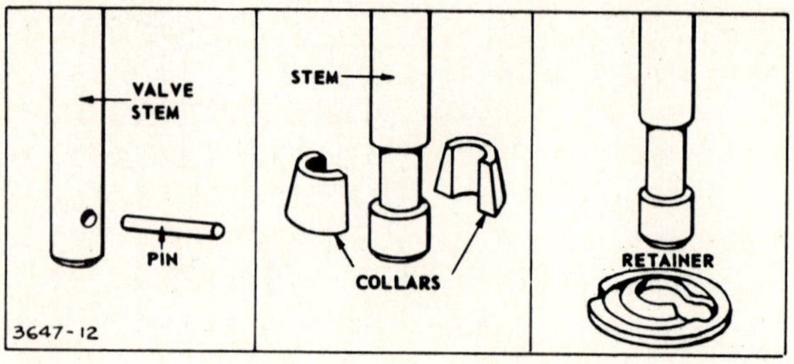

FIGURE 4-15. Variation in valve spring retainers (Courtesy of Briggs & Stratton Corporation).

valve lifters (described next) to open the appropriate valve at the appropriate time. Valves are closed, as noted earlier, by the valve springs. (See Figures 4-16 and 4-17.)

The shape of the valve lobes determines how long the valves remain open and closed. However, exactly when the valves open, and how often

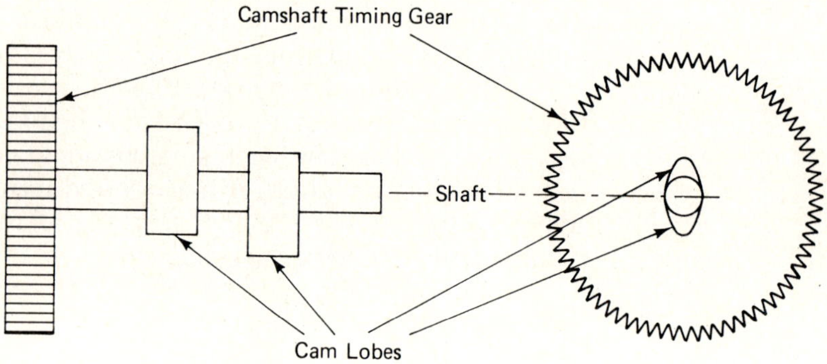

FIGURE 4-16. Camshaft schematic.

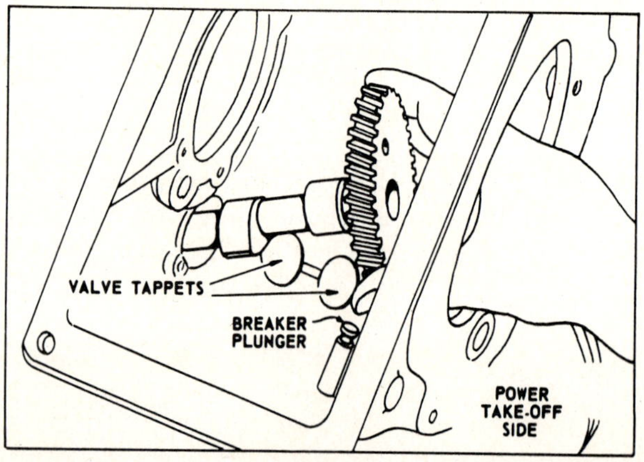

FIGURE 4-17. Installing the cam into the crankcase (Courtesy of Briggs & Stratton Corporation).

they open and close are controlled by the relationship between a driven gear on the end of the camshaft and a driving gear on the end of the crankshaft.

The rate of opening and closing depends on the number of teeth in each gear. Since each valve must open once on every other engine revolution, the camshaft must rotate at half the crankshaft speed. Therefore, the camshaft gear has twice as many teeth as the crankshaft gear. (See Figure 4-18.)

To ensure that the valves open at the right time during each of the four strokes, the camshaft and crankshaft must rotate in exactly the correct relationship. In other words, the camshaft lobes must be turned in a certain way when the crankshaft is in a certain position. Correct valve timing is obtained during engine assembly by matching alignment marks on the crankshaft and camshaft.

The camshaft in most small engines is cast from steel or iron and then machined to final tolerances. On smaller, less expensive engines, the timing gear might be cast integrally with the camshaft. The crankshaft timing gear is usually fabricated separately, then held in position with a key.

FIGURE 4-18. **Bottom view of a small engine, showing the camshaft and crankshaft timing gears (Courtesy of Briggs & Stratton Corporation).**

Valve Lifters

Valve lifters (or tappets) transmit the action of the camshaft lobes to the valve stems. Located in holes drilled in the block, one end of the lifter rests on the camshaft lobe; the other end is positioned directly beneath the valve stem. Because of expansion and contraction caused by changes in engine temperature, a small clearance is usually provided between the top of the lifter and the valve stem. (See Figure 4-19.)

Valve Train Variations

Most small engine valve train components, such as those just described, are located entirely in the block. However, a few small engines (like most automotive and motorcycle engines) locate the ports and valves in the heads.

FIGURE 4-19. Valve system components: adjustable tappets, intake valve, exhaust valve, springs, spring retainer, and retainer locks (Courtesy of Onan).

Push rods extend up from the camshaft, through the block and head. The push rods engage rocker arms located on top of the head (under a valve cover). When one end of a rocker arm is lifted up by a push rod (as the camshaft lobe pushes the rod up), the rocker arm "rocks" down on the end of the valve stem to open the valve. A valve spring closes the valve and causes the rocker arm to "rock" in the other direction after the cam rotates sufficiently. Except for these differences, "overhead valve" engines operate the same way as the typical "flathead" engine described in this chapter. (See Figure 4-20.)

There is one other variation even less common than the overhead valve configuration. It is called the "overhead camshaft" or "cam-in-head" configuration. As the name implies, the camshaft is located (or attached to) the head. Driven from the crankshaft by a timing belt or chain, the camshaft lobes directly engage the valve stems. (See Figure 4-21.)

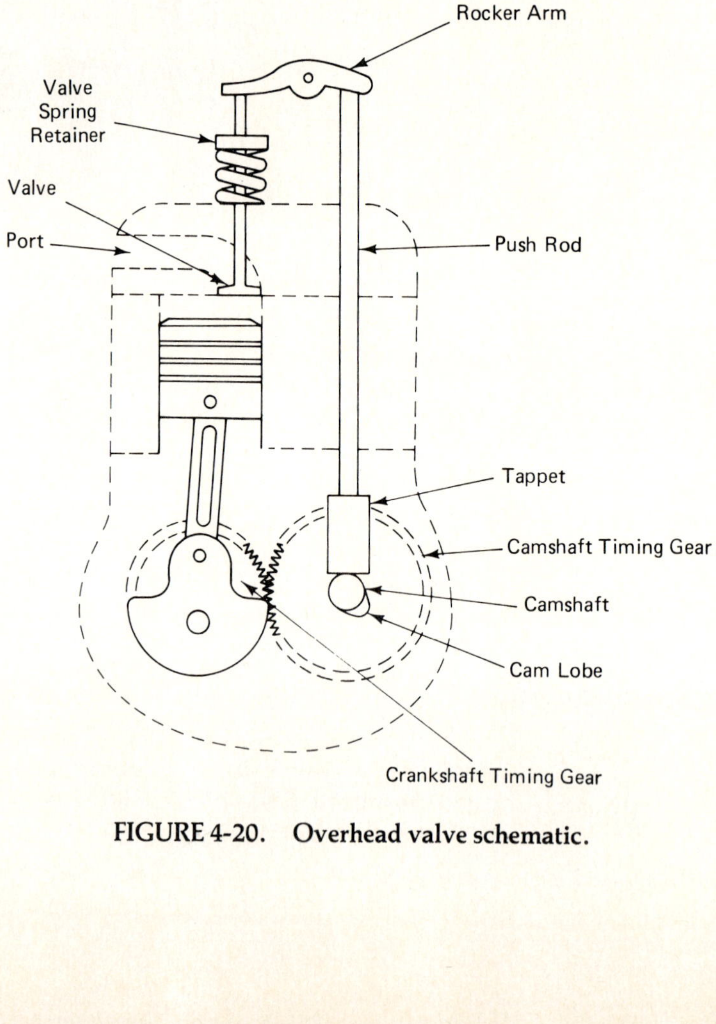

FIGURE 4-20. Overhead valve schematic.

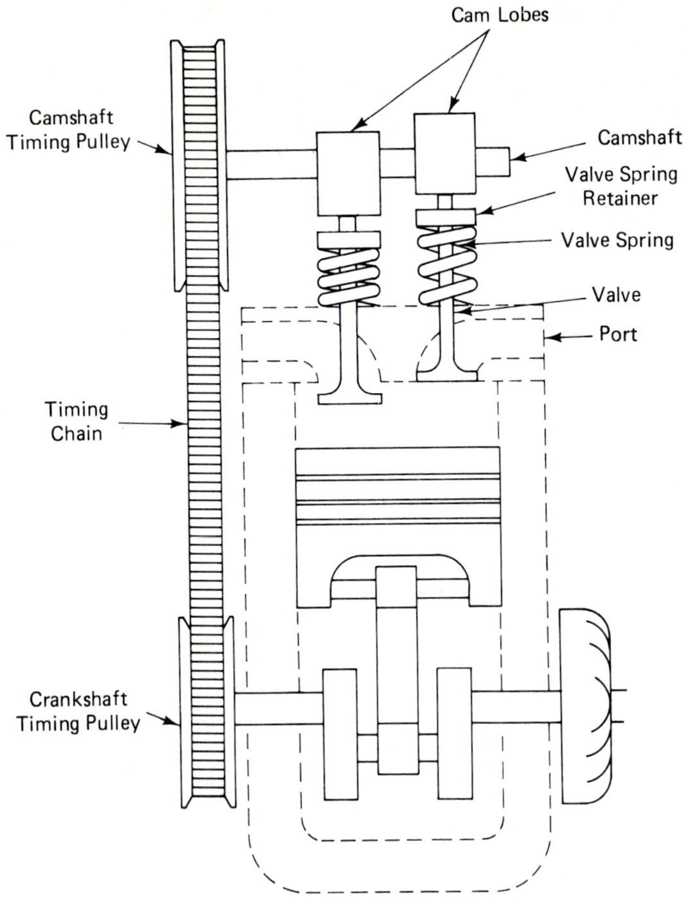

FIGURE 4-21. Overhead cam schematic.

HEAD

The last item that we will discuss is the head. It seals the top of the block and, along with the cylinder and top of the piston, defines the combustion chamber. Most small engines are of the "flathead" variety. This means, as noted above, that the head does not contain any of the valve train components. However, the underside of the head is shaped to provide opening clearance for the valves. The shape also helps to promote a swirling action of the air/fuel mixture, which, in turn, ensures complete mixing of the air and fuel. Also, the cavity on the underside of the head helps define the compression ratio. (See Figure 4-22.)

The head, like the block, may be cast from aluminum or iron. Like the block, the top surface of the head is finned to dissipate heat. The head is positioned on the block by studs projecting from the mating surface of the block through matching holes drilled in the head. Nuts screwed onto the studs in a particular tightening sequence hold the head in place. Sealing between the block and the head is provided by a special heat-resistant, high-pressure gasket. (See Figure 4-23.)

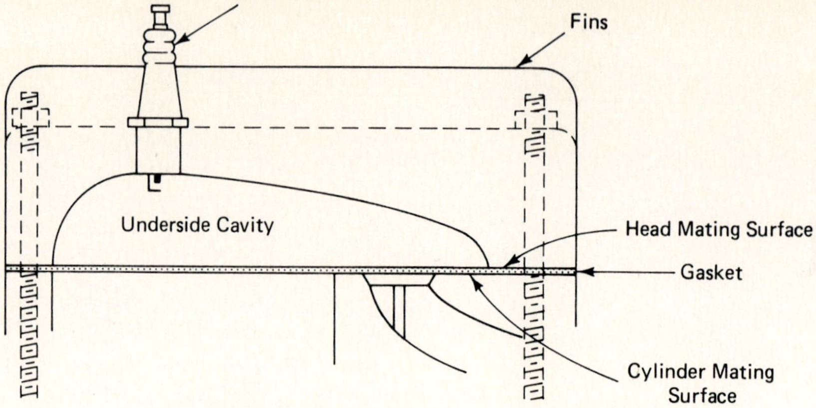

FIGURE 4-22. Cylinder head schematic.

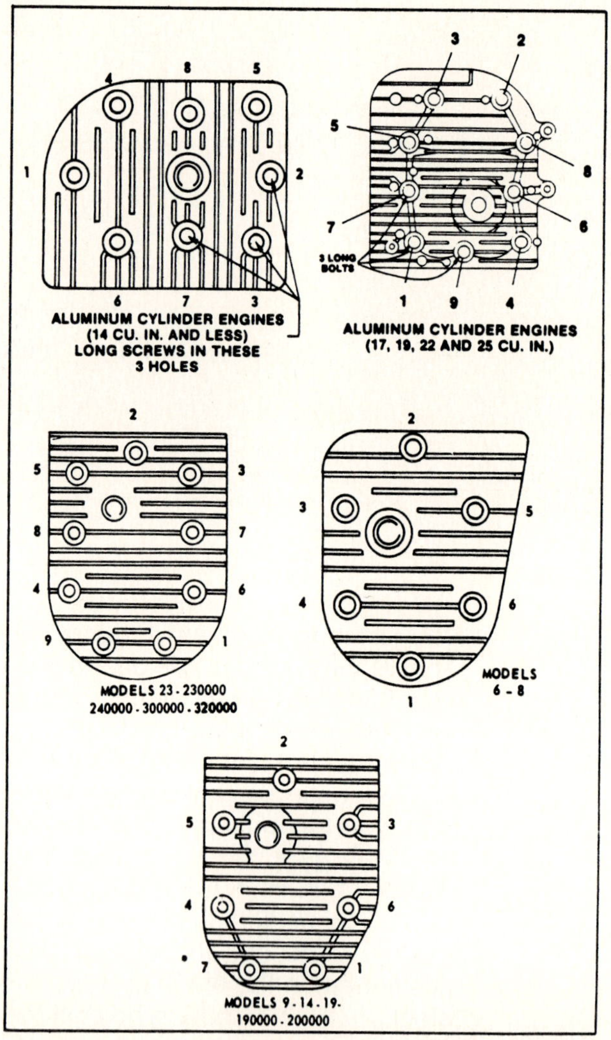

FIGURE 4-23. Several types of cylinder heads (Courtesy of Briggs & Stratton Corporation).

5

FUEL SYSTEMS

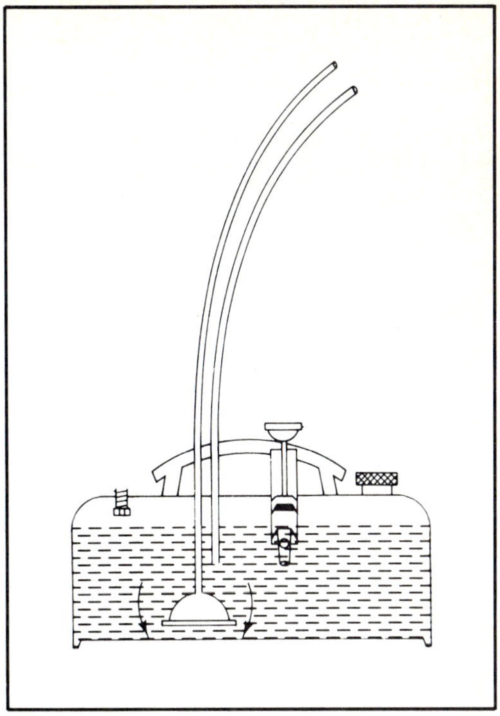

GENERAL

This chapter starts out by reviewing the major types of fuels used in small gasoline engines. Then the components and operation of some typical gasoline engine fuel systems are examined.

TYPES OF FUEL

Gasoline

Most small engines burn gasoline. It is the most common of all fuels. It is also relatively easy to handle, does not require any special operating procedures, and is able to take advantage of an established, widespread technology. Furthermore, though all fuels are expensive in this age of scarce energy resources, gasoline prices are competitive with the cost of most other fuels. (See Figure 5-1.)

"Regular" leaded or unleaded gasoline with a median octane rating of about 90 is most often specified by small engine manufacturers.

The biggest problem with gasoline is its high volatility. Gasoline readily evaporates into an explosive vapor state. Therefore, small engines that use gasoline should never be stored in closed spaces or around heat or flames. Dangerous levels of fuel vapor might be produced in the tank or carburetor float bowl. The same caution holds true for gasoline stored in any sort of container. Check with your fire department for local regulations regarding gasoline storage.

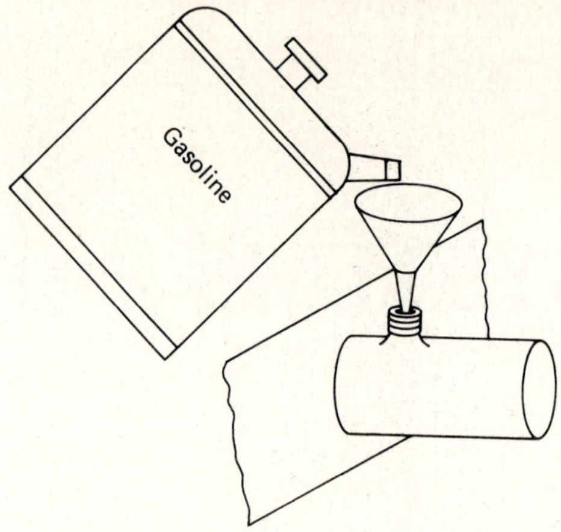

FIGURE 5-1. Gasoline can, funnel, and tank.

Kerosene

Some heavy duty industrial small engines use kerosene. In many cases, these engines employ gasoline for starting, then switch over to kerosene after the engine becomes hot. (See Figure 5-2.) That is because kerosene doesn't vaporize as readily as gasoline, and consequently, doesn't ignite as easily. Such engines are equipped with specially adapted carburetors that have heat radiating surfaces built in or located nearby. The compression ratio is also likely to be lower than in a comparable "pure" gasoline engine. Early farm tractors employed this principle, probably because kerosene or heating oil was more available than gasoline.

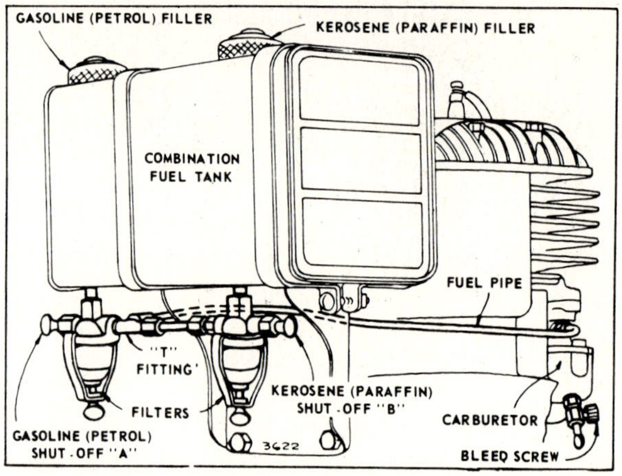

FIGURE 5-2. Combination fuel tank used for gasoline and kerosene (Courtesy of Briggs & Stratton Corporation).

Natural Gas

A few small engines in stationary applications use natural gas—virtually the same sort of gas that is piped into homes for heating. One advantage of such engines is the simplicity of the fuel delivery and mixing systems. Since the gas is delivered under pressure, no fuel pump is required. Also, because the gas turns into an easily ignited vapor when the pressure is reduced, there is no need for a complicated carburetor. All that is required is a mixing valve and several regulators. Of course, the major disadvantage is immobility; these engines must remain connected to a delivery pipe or service line. Reduced power is another disadvantage. Compared to gasoline, natural gas produces 15 to 20 percent less horsepower in similar engines.

Liquid Petroleum (LP) Gas

So called "bottled gas" (either propane or butane) is used by some small engines. Like natural gas, pressurized LP gas does not require complicated fuel delivery or mixing systems. However, unlike natural gas, LP gas "bottles" can be transported, which means that engines using this fuel are portable. Also, unlike natural gas, LP gas does not suffer a horsepower loss when compared to gasoline; similar LP- and gasoline-powered engines produce about the same power.

Liquid petroleum-powered engines are most often used indoors or in poorly ventilated areas. That is because the burning process does not produce poisonous exhaust. Gasoline, on the other hand, yields carbon monoxide, an odorless, colorless, and deadly gas. One major disadvantage of LP gas compared to gasoline is the relative size and weight of the fuel storage containers. To provide comparable energy reserves, an LP gas bottle must be much larger and heavier than a gasoline fuel tank. (See Figure 5-3.)

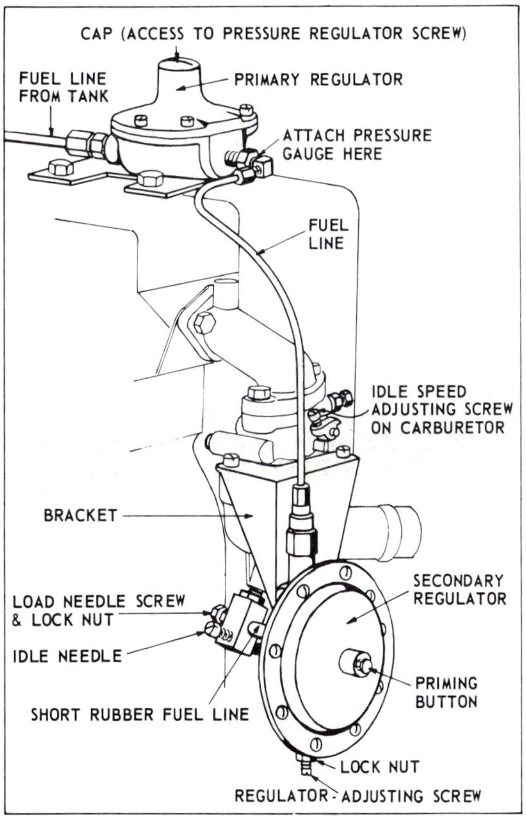

FIGURE 5-3. Regulators, lines, and equipment for LPG system (Courtesy of Briggs & Stratton Corporation).

Diesel Fuels

A few small engines use diesel fuel. However, unlike the fuels described previously, diesel fuel makes some strict demands on starting and fuel delivery systems and on engine design. Here is the reason: Diesel engines do not use a spark plug to ignite the fuel mixture. Instead, the fuel is injected directly into the cylinders during the last part of the compression stroke.

Hot air, squeezed between the piston and the cylinder head, starts the burning process. For diesels to work properly, the fuel must then be injected at very high pressures at just the right instant. This requires a sophisticated, often complex fuel injection system. Also, in order to squeeze the air into a sufficiently dense (and therefore, hot) state, diesel engines must have very high compression ratios. Usually, the compression ratio of a diesel engine is at least twice that of a comparable gasoline-powered engine. To operate safely and reliably at these elevated compression ratios, diesels must be stronger and heavier than gasoline engines.

The advantages most often cited for diesels are reduced maintenance and improved fuel economy. Usually, for small engines, these advantages do not outweigh the added cost, weight, and complexity. However, there is another diesel advantage that did prompt the Army at one time to use diesels to operate small electric generators. Diesel fuel is less volatile than gasoline, which means it presents less of a fire and explosion hazard when stored for long periods (as might be the case for infrequently used generators).

Two-Cycle Oil/Gasoline Mixture

Small two-cycle engines use fuel for a special purpose. Oil is mixed with the gasoline to lubricate the crankcase and cylinder area. (Remember: the air/fuel mixture goes from the carburetor to the crankcase and then to the combustion chamber).

Because the oil and gasoline are mixed, two cycles receive the same lubrication, no matter how the engine is turned. It is for this reason that two-cycle engines are often employed in applications where the engine might end up in odd positions. (Which explains why most chain saws are powered by two-cycle engines.)

The exact amount of oil mixed with gasoline depends on the manufacturer's specifications. Sixteen parts of regular gasoline to one part of special two-cycle lubricating oil is a typical mixture. Special care must be taken to obtain the correct proportions. Add too much oil, and the engine is likely to become gummy with deposits. Add too little oil, and there is the risk of metal-to-metal contact and subsequent engine damage or outright failure.

BASIC FUEL SYSTEM FUNCTIONS

All of the fuels described in the preceding section represent stored energy. When they are mixed with air and exposed to heat, the energy is released in the form of hot, expanding gases. Consequently, the major functions of any fuel system are:

1. Storing fuel

2. Filtering fuel
3. Delivering fuel
4. Mixing air and fuel

The rest of this chapter describes the components and operations required to perform these functions in small gasoline-powered engines.

FUEL TANKS FOR STORING FUEL

Fuel tanks for small gasoline engines come in all shapes and sizes. Some are made from sheet metal. Others are plastic. Capacities likewise vary. The tanks found on small, hand-held two cycles might not hold more than a pint or so, whereas the tanks used in garden tractors might hold several gallons. These are about the only common features:

1. The filler cap is located on top.
2. The fuel outlet is located on the bottom.
3. Because some small engines use gravity fuel feed, the tank may be located higher on the engine than the carburetor.

FUEL LINES

Many early small engines used steel fuel lines with brass fittings threaded into the tank outlet at one end and into the carburetor inlet at the other. Most modern small engines use synthetic rubber lines and plastic fittings. They seem to work about as well and are less expensive.

FILTERING FUEL

An important function in any engine, large or small, is to make sure the fuel is clean. This is especially true for small engines that might be operated in dirty, hostile environments.

The simplest type of filter is a wire mesh located between the tank outlet and the fuel line. When the filter becomes clogged with trash, it is removed and cleaned. (See Figure 5-4.)

The sediment bowl is a more sophisticated filter. It contains a porous ceramic element inside a bowl. Fuel flows into the bowl, through the filter, and into the fuel line leading to the carburetor. Sediment trapped by the

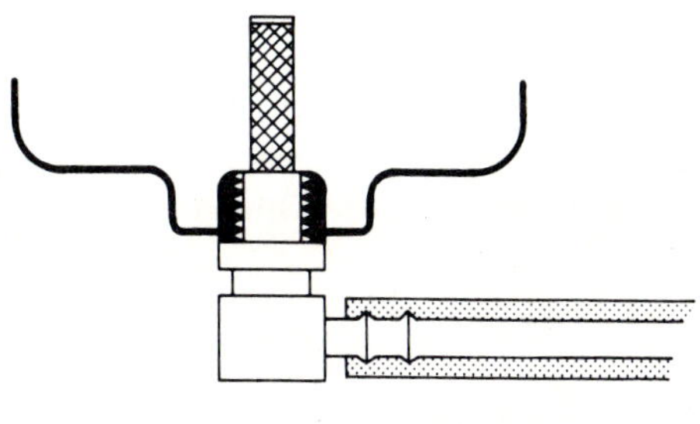

FIGURE 5-4. Screen-type tank filter.

ceramic element drops to the bottom of the bowl. Water contained in the fuel also settles to the bottom of the bowl.

All sediment bowls are detachable and many are made of glass so that the amount of trapped debris can be visually checked. When the bowl becomes full, it is simply removed, cleaned out, and then reattached to the filter body. (See Figure 5-5.)

Another very common filter is the so-called in-line type. These filters are tube-shaped, with an inlet on one end and an outlet on the other. Located inside the clear plastic shell is a fine mesh paper filter. When the filter becomes clogged, it is replaced. (See Figure 5-6.)

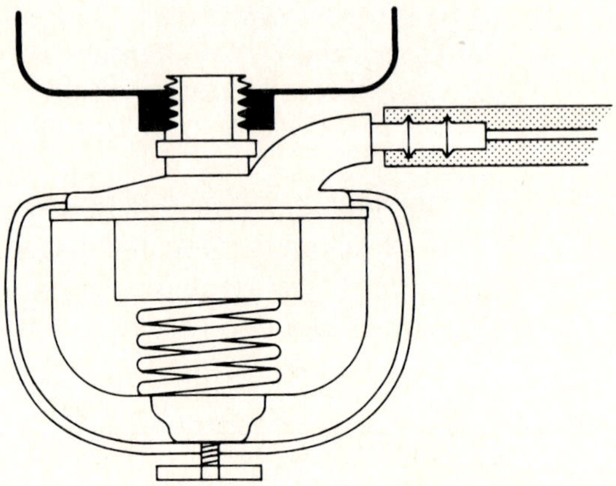

FIGURE 5-5. A sediment-bowl filter.

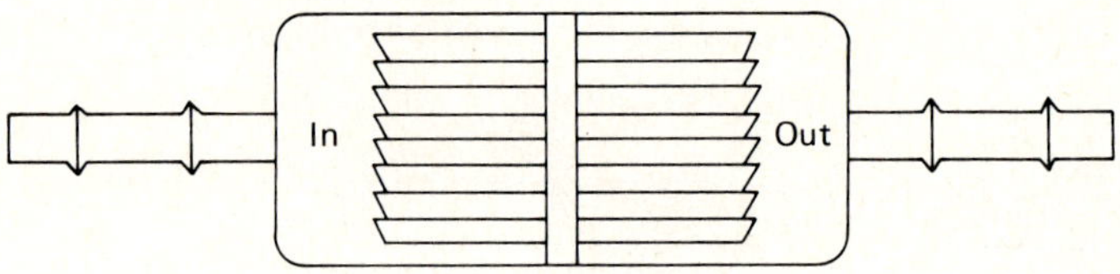

FIGURE 5-6. An in-line filter.

FUEL DELIVERY

Many small engines rely on the force of gravity to deliver fuel from the tank to the carburetor. However, in some cases, a gravity feed system is not practical. Perhaps the engine will be turned in a position so that the tank is no longer above the carburetor, or perhaps the design simply doesn't allow the gas tank to be placed over the carburetor. In either case, it is necessary to force the fuel from the tank to the carburetor. Here are some of the methods that have been used to deliver fuel in small gasoline engines.

Mechanical Fuel Pumps

Mechanical fuel pumps are used in many small gasoline engines. They are similar in design to the mechanical fuel pumps used in automobiles. (See Figure 5-7.) The main components include an aluminum housing with one inlet valve, one outlet valve, and a rubberized diaphragm. The pump is

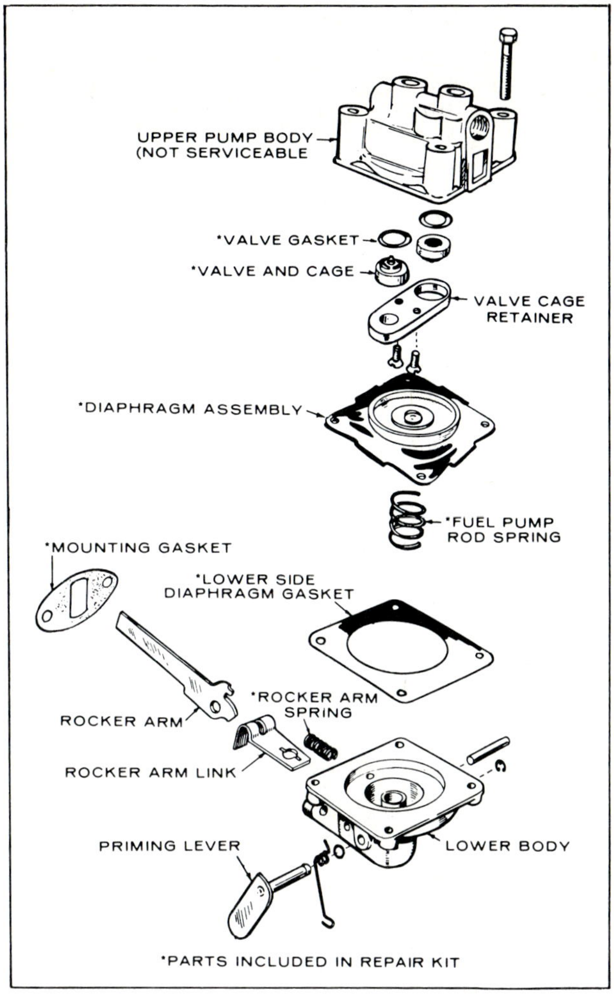

FIGURE 5-7. Cross-sectional exploded view of mechanical fuel pump (Courtesy of Onan).

mounted on the engine casting. An actuating lever or arm protrudes from the pump into the engine, where it rides on an eccentric surface machined onto the camshaft. The rotation of the camshaft causes the arm to rise and fall, which in turn causes the diaphragm to go up and down (actually, the diaphragm is pulled down by the arm and pushed up by the diaphragm spring). The motion of the diaphragm "pulls" fuel through the one-way inlet valve and "pushes" fuel through the one-way outlet valve. Fuel pressure is regulated between 3 and 6 pounds per square inch by the tension of the diaphragm spring.

Vacuum/Pressure Operated Fuel Pump

Some small gasoline engines (either two or four strokes) use the vacuum produced by the engine to operate the fuel pump. Such pumps employ a diaphragm and work in much the same way as a mechanical pump. However, instead of moving the diaphragm with a cam actuated lever, engine vacuum is applied to one side of the sealed diaphragm. Since there is atmospheric pressure in the vented fuel tank, fuel is forced into the lower pressure region created on the other side of the diaphragm. As in mechanical systems, a spring pushes the diaphragm in the opposite direction when vacuum is no longer applied. Also, one-way check valves make sure that fuel flows in the proper direction through the inlet and outlet ports. (See Figure 5-8.)

In four-cycle engines, vacuum is applied when the engine is in the intake stroke. In two-cycle engines, vacuum is supplied from the crankcase when the piston is on the upstroke.

Some pumps employ a variation of the crankcase system. Instead of using vacuum produced in the crankcase when the piston is on its upstroke, pressure produced by the opposite piston stroke is applied to one side of the diaphragm. The result is the same: The diaphragm moves and fuel is delivered from the tank to the carburetor.

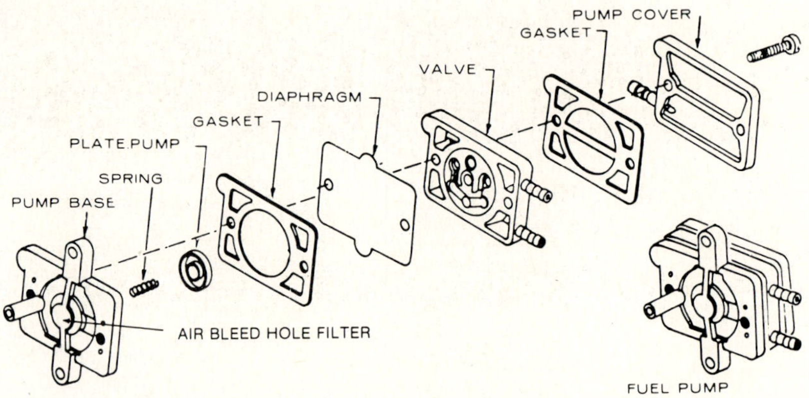

FIGURE 5-8. Exploded view of a fuel pump (Courtesy of Onan).

Pressurized Systems

These systems are common on small outboard marine engines. The tank is usually a portable metal device with a carrying handle which allows the tank to be carried to a fill-up point. When the tank contains the proper mix (in the case of two cycles), it is placed in a convenient location and connected to the engine by two rubber hoses. These hoses are usually molded into one piece and attached by "quick connect" adapters to the engine and the tank. One line connects the tank to a valve leading into the crankcase. When the engine is running, pressure from the crankcase forces fuel from the tank, through the second delivery line, to the carburetor. Initially, when the tank is first filled, pressure for starting the engine is obtained by a hand-operated pump, usually located in the tank carrying handle. (See Figure 5-9.)

FUEL MIXING SYSTEMS

As we have already noted, fuel must be mixed with air in order to burn. However, just any sort of mixing will not do. The liquid fuel must be turned into a gaseous vapor so that the air (also a gas) and fuel are thoroughly

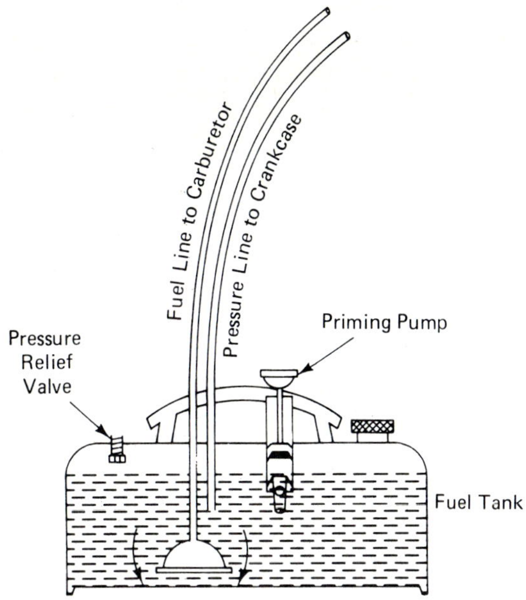

FIGURE 5-9. Typical pressure fuel tank.

mixed. Nor will just any amount of air and fuel work. For best results in a gasoline-powered engine, there should be about 14.7 parts of gaseous fuel for each part of air.

The function of the fuel mixing system is to convert the proper amount of liquid fuel into a vapor and to mix the vapor with air. Before we get into the operation of fuel mixing systems, it will be helpful to briefly discuss some basic principles underlying the behavior of liquids and gases.

UNDERLYING CONCEPTS

Vaporization

Several factors affect the way fuels like gasoline change from a liquid to a gaseous state.

TEMPERATURE

Warmer liquids evaporate or vaporize faster than cooler liquids. That is because the tiny molecules which make up the liquid move faster when they are warmer. Vaporization occurs when a molecule moves fast enough to escape from the exposed surface of the liquid into the atmosphere.

SURFACE AREA

All transition from a liquid to a gaseous state occurs at the exposed surface of the liquid. Therefore, the more surface that is exposed, the faster the liquid becomes a vapor. To illustrate this point, imagine that you have placed equal amounts of gasoline in two containers in the middle of an empty parking lot. Also, imagine that one container is wide and shallow and one is tall with a small opening. If you left for several hours, upon returning, you would discover that more gasoline had evaporated from the wide shallow container. Why? Because it exposed more liquid surface to the atmosphere; consequently, more molecules were able to escape. (See Figure 5-10.)

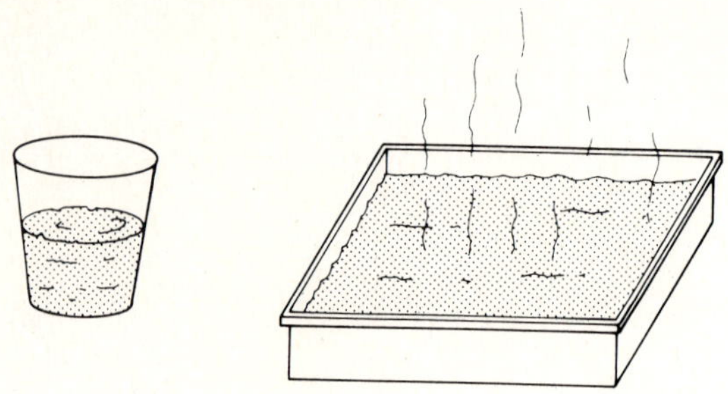

FIGURE 5-10. Demonstrating evaporation with a pan and tumbler.

PRESSURE

Another factor affecting vaporization is the pressure of the atmosphere adjacent to the liquid surface. When the atmospheric pressure is relatively high, its molecules are more densely packed. This makes it more difficult for the fuel's molecules to escape from the liquid surface. However, when the pressure is lower, the air molecules are further apart, which allows more fuel molecules to escape.

To illustrate this action, consider the operation of a rubber bulb atomizer (the kind once used to spray perfume). (See Figure 5-11.) As the bulb is squeezed, air flows over a pick-up tube which extends down into the liquid. The molecules of moving air are separated more than in stationary air, so the pressure over the pick-up is reduced. Normal atmospheric pressure adjacent to the liquid surface pushes the liquid up into the tube. Upon entering the low-pressure region, the liquid is turned into tiny droplets which vaporize more readily because of the reduced pressure (also because the liquid is broken into numerous droplets, each exposing a complete spherical surface where vaporization can take place).

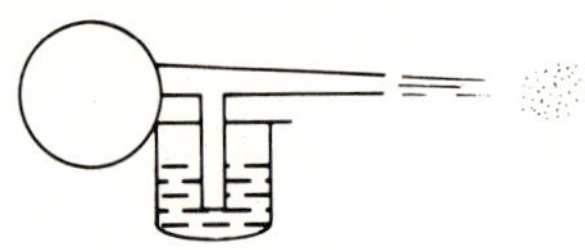

FIGURE 5-11. An atomizer.

Differential Pressure Fuel and Air Delivery

In the previous illustration, liquid was delivered because there was a difference in pressure between the stationary air adjacent to the liquid and the moving air over the pick-up tube. The same basic principle is used to deliver fuel and air in internal combustion engines. In other words, air and fuel are pushed by normal atmospheric pressure into low-pressure regions created inside the engine or carburetor.

The two primary regions of reduced pressure are in the cylinder and carburetor throat. Reduced pressure is created in the cylinder during the intake stroke. As the piston moves away from TDC, the cylinder volume expands. Consequently, the air in the cylinder also expands to fill the increased area. This causes the pressure to drop (because the air molecules are not so tightly packed). Air at normal pressure outside the engine can then push down the carburetor throat through the intake port and into the cylinder's low-pressure region.

The second low-pressure region is created by the movement of air going through the carburetor throat to the cylinder. As we saw in the squeeze bulb example, fast-moving air has lower pressure than stationary air. Therefore, the pressure in the carburetor throat is reduced. If fuel is stored in a region of atmospheric (or greater) pressure, and if delivery passages are provided to the carburetor throat, the fuel will be pushed through the passages to the lower-pressure region. (See Figure 5-12.)

In many cases, the pressure in the carburetor is reduced even further by forcing moving air to pass through a restriction in the throat. This restriction is known as the *venturi*. The differential pressure causing the air to move through the throat will attempt to move the same amount of air, regardless of the restriction. Therefore, the air will speed up as it moves through the venturi. As we've seen before, the faster the air moves, the more the pressure is reduced.

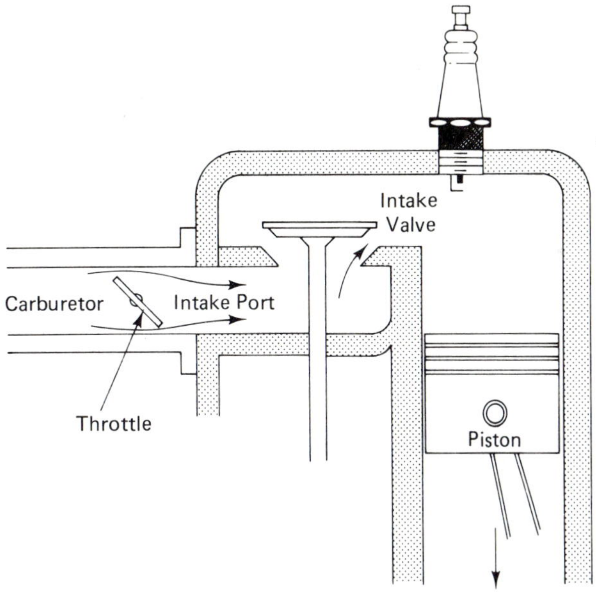

FIGURE 5-12. Sketch of a cylinder showing the fuel intake system.

CARBURETORS

Now that we have examined the basic concepts underlying the operation of fuel mixing systems, let's look at the components and operation of the most common fuel mixing device, the carburetor. (See Figure 5-13.)

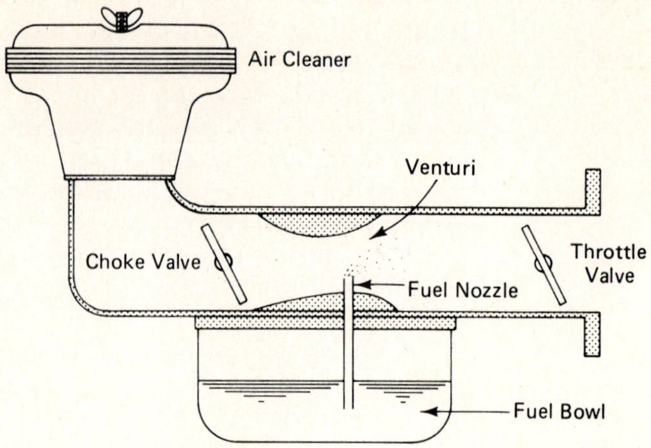

FIGURE 5-13. The basic carburetor showing the fuel bowl and nozzle.

Basic Carburetor Components

All small engine carburetors have a number of components in common:

1. *Throat/venturi.* For passing air from outside the engine to the intake port. Also creates a low-pressure region to help deliver fuel into the air stream.
2. *Temporary fuel storage.* For holding a small quantity of fuel for delivery into the air stream.
3. *Fuel flow regulator.* For controlling the amount of fuel entering the storage area and for helping to control the flow from the storage area.
4. *Air flow regulator.* For controlling the amount of air entering the engine, thereby controlling engine speed. Most often the throttle plate is located in the base of the throat. (Open the throttle plate, more air flows, and the engine speeds up. Close the throttle plate, and the engine slows down.)
5. *Air cleaner.* For filtering small particles out of the air before it enters the carburetor throat.
6. *Idle speed circuit.* For delivering fuel when the engine is operating at reduced or idle speeds.
7. *Main delivery circuit.* For delivering fuel when the engine is running faster.
8. *Choke.* For providing additional fuel when the engine is cold and the fuel does not vaporize as readily.

The exact appearance of these components and their detailed operation depends on the particular small engine application. Some engines must operate at a variety of speeds over a range of conditions. Other engines run at fixed speeds in a constant environment. Other engines are stationary, while others are mounted on tractors or similar kinds of moving vehicles. Some so-called hand-held engines may be operated in *any* position, even upside down.

The next paragraphs examine a few of the basic types of carburetor system used to satisfy the demands imposed by these various applications.

AIR CLEANERS

All carburetors require an air cleaner of some kind. Dust and debris must be strained from the air before it enters the engine. This is especially important for most small engines, since they are likely to be operated in very dirty conditions. Various kinds of filters are used in small engine air cleaners—paper elements, oil baths, foram elements soaked in oil, etc. Figures 5-14 through 5-18 illustrate some typical examples.

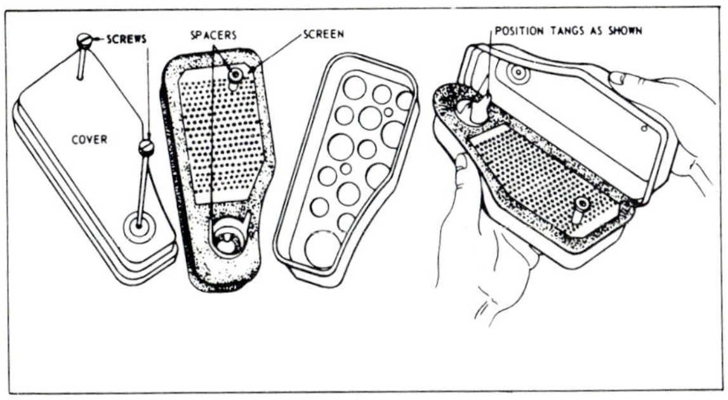

FIGURE 5-14. Standard air cleaner (Courtesy of Briggs & Stratton Corporation).

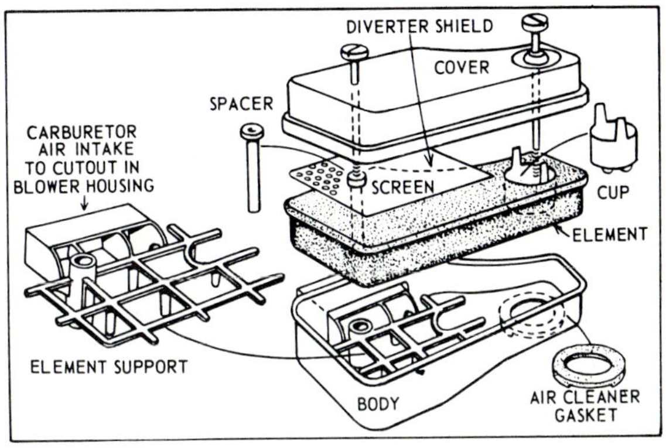

FIGURE 5-15. Ducted air cleaner (Courtesy of Briggs & Stratton Corporation).

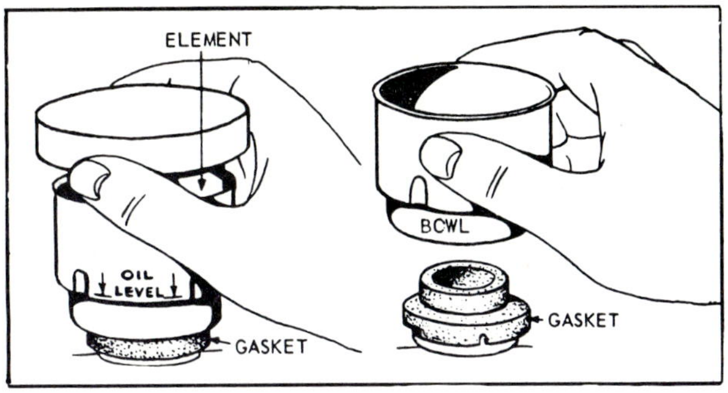

FIGURE 5-16. An oil bath air cleaner (Courtesy of Briggs & Stratton Corporation).

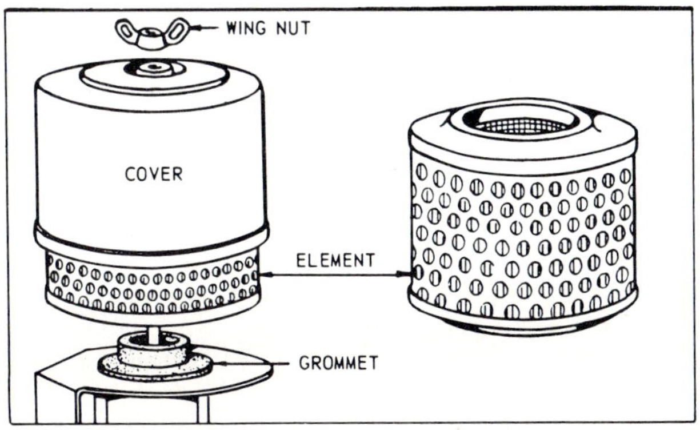

FIGURE 5-17. A dry element air cleaner (Courtesy of Briggs & Stratton Corporation).

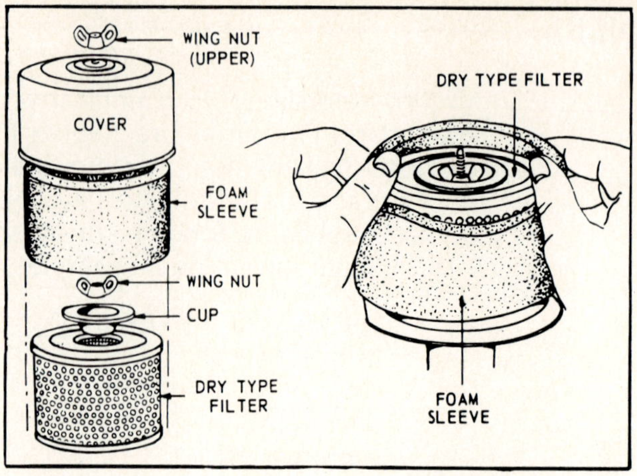

FIGURE 5-18. A heavy-duty air cleaner (Courtesy of Briggs & Stratton Corporation).

FLOAT BOWL CARBURETOR

This type of carburetor is common on many small engines (except those for hand-held applications). As the name implies, the characteristic feature of this carburetor is a float bowl. The float and associated components control the flow of fuel into the carburetor throat and provide a temporary fuel storage area. (See Figure 5-19.)

Fuel comes to the bowl from a fuel pump or from a gravity feed tank. Passage into the bowl is controlled by a needle valve and seat. The valve is connected to the float, which rides on the gasoline contained in the bowl. As the fuel level increases, the float rises, which pushes the needle up into the seat and closes the valve. When the fuel is withdrawn and the float drops, the valve lets more fuel into the bowl.

The maximum level of fuel in the bowl is controlled by the uppermost position of the float, which is adjusted by bending the float arm up or down.

Fuel is pushed out of the vented bowl by differential pressures be-

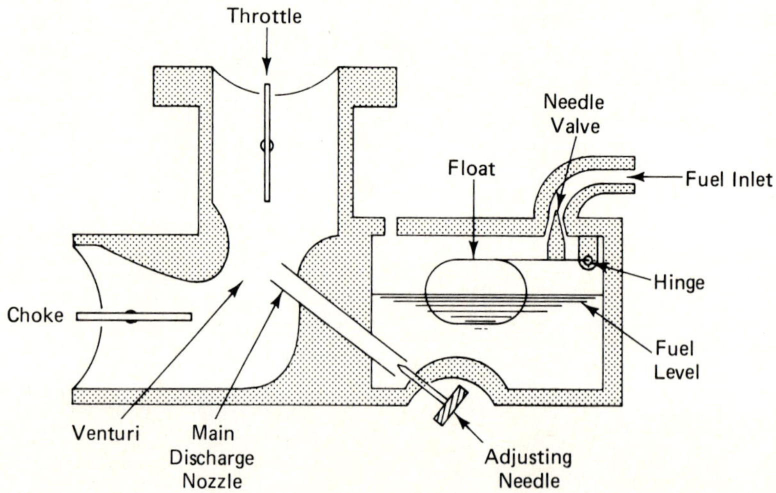

FIGURE 5-19. The float system of a typical carburetor.

tween the air in the bowl and the low-pressure regions developed in the carburetor. Exactly where the fuel goes depends on the throttle setting and the resultant engine speed. Figure 5-19 shows the two main flow circuits.

Main Discharge Circuit

The main delivery passage extends from an opening in the float bowl up to the venturi region of the carburetor throat. During normal engine operation (from one-third to full throttle) pressure in the venturi is reduced enough to allow normal atmospheric pressure to push the fuel out of the main discharge nozzle. As the throttle opens, pressure is reduced accordingly, which means that proportionally more fuel will be delivered into the air stream, thus maintaining a nearly constant air fuel mixture.

Idle Speed Circuit

When the throttle moves toward the closed position, air flow through the venturi is reduced. Past a certain point, the pressure differential will not be enough to force fuel from the main discharge nozzle. Therefore, at idle and reduced engine speeds, another source of reduced pressure must be provided.

A typical idle speed circuit is shown in Figure 5-20. The low-pressure region is created on the side of the throttle plate nearest the engine. When the throttle is closed or nearly closed, the partial vacuum in the cylinder extends up to the throttle plate. A passage leads from the low-pressure region into a tube that extends across the venturi through the main discharge nozzle into the float bowl. When the pressure beneath the throttle plate is less than the pressure at the main discharge nozzle, fuel flows through the idle passage.

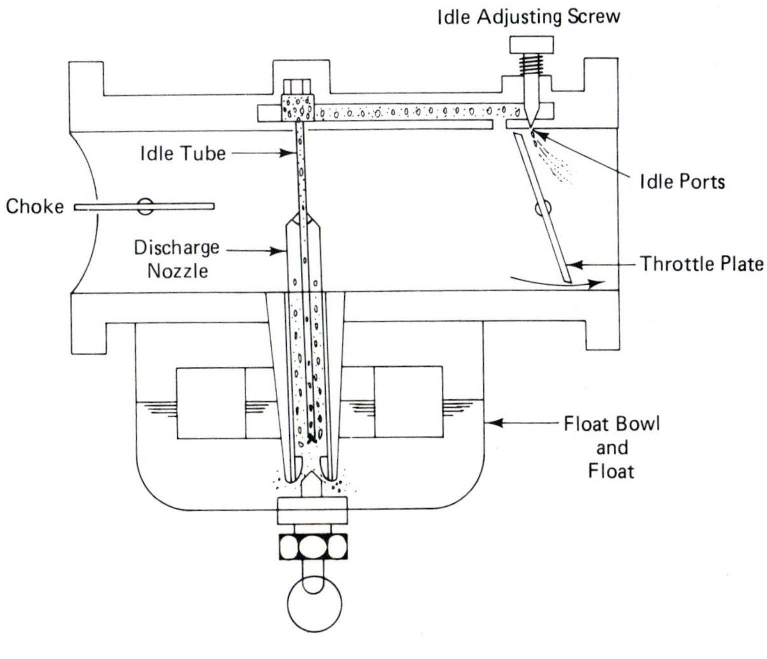

FIGURE 5-20. A basic carburetor idle circuit.

To aid in vaporization, several air vents or "bleed" holes are provided in the idle circuit passage. Air coming into these holes mixes with the fuel to create a mass of bubbles and droplets. That way the fuel's surface area is increased, which causes vaporization to proceed faster.

Choke Circuit

Some means is usually provided to help start cold engines. The fuel doesn't vaporize as readily when it is cold, which could cause the mixture to be too "lean" (containing too little fuel).

The device most often used in float bowl (and other types of carburetors) is a choke plate located at the entrance to the carburetor throat (called the air horn). When the choke plate is closed, vacuum from the cylinder extends all the way to the top of the carburetor. Responding to this pressure reduction, normal atmospheric pressure pushes fuel through the main discharge and idle circuits. (See Figures 5-19 and 5-20.)

The choke in small engines is almost always manually controlled. It is up to the operator to open the choke as the engine warms up. Failure to do so will cause the mixture to be too rich (contain too much fuel). Raw gasoline may enter the cylinder and wash the oil off the cylinder walls. Gasoline might also run into the crankcase and dilute the main oil supply.

Carburetor Adjustments

The type of adjustments provided for float-bowl-type carburetors depends on the particular application. Some may have no adjustments at all. However, most provide: (1) a screw-type stop to control the position of the throttle plate at idle speeds, and (2) a screw-positioned needle valve to adjust the amount of fuel flowing through the idle circuit (and hence to control the mixture at idle speeds). Some float bowl carburetors also have a screw-positioned needle valve to control the amount of fuel flowing from the main delivery outlet in the float bowl. Such an adjustment determines the fuel mixture during normal operating speeds. (See Figures 5-19 and 5-20.)

CARBURETORS FOR HAND-HELD ENGINES

Several features of float-bowl-type carburetors are common on all types of small engine carburetors. For instance, most employ differential pressures for delivering fuel from a temporary storage area into a low-pressure venturi region. Most also use a throttle plate to control air flow and engine speed. Most have a choke system to block air flow into the air horn and thereby enrich the mixture for cold starting. Many have separate idle and main fuel delivery circuits. (See Figure 5-21.)

However, there is one significant difference between float-bowl-type carburetors and those used on hand-held engines: namely, the method used to control the amount of fuel maintained in the temporary storage area. Float-bowl regulating systems will work only if the float is in a more or less horizontal position. Turn the engine sufficiently, and the force of gravity will no longer cause the float to move in the correct direction.

A fuel injection system could be used to avoid the problem; how-

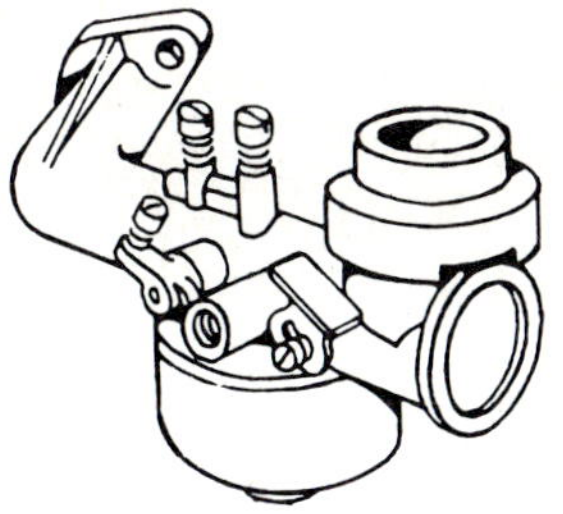

FLO-JET ONE PIECE

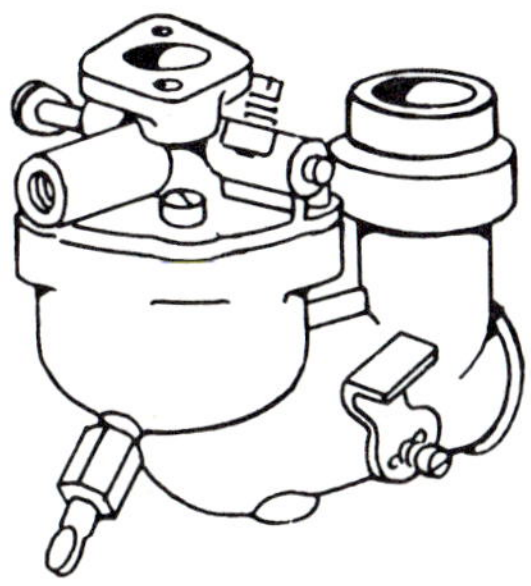

FLO-JET TWO PIECE

FIGURE 5-21. One- and two-piece designs for flow-jets (Briggs & Stratton Corporation).

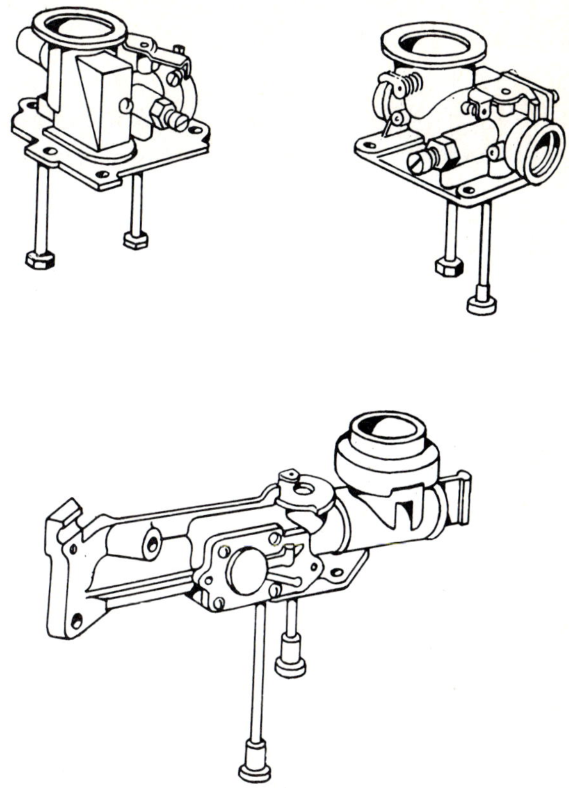

FIGURE 5-22. Pulsa-jet carburetors (Courtesy of Briggs & Stratton Corporation).

ever, that would be too expensive. The most common solution is to pressurize the fuel in the temporary storage area and control the flow of fuel with a spring-loaded diaphragm. Here are two examples.

Pulsa Jet

The Pulsa Jet system employs a diaphragm-type pump built into the carburetor. (See Figure 5-22.) The diaphragm is vacuum operated—one side is connected by passages to the intake port. Every time an intake stroke occurs, vacuum from the cylinder pulls the diaphragm against a calibrated spring and draws fuel into the storage area. Then when the intake stroke stops, the spring pushes the diaphragm back the other way. The amount of fuel entering the storage area is regulated by the spring tension and the amount withdrawn by the venturi. As a result, the storage area is always full, regardless of the position of the engine. (See Figures 5-23 and 5-24.)

Diaphragm Carburetor

This type of carburetor is common on many types of hand-held engines. It features a fuel storage area sealed on one side by a spring-loaded diaphragm. (See Figure 5-25.) Fuel is supplied to the storage area from a

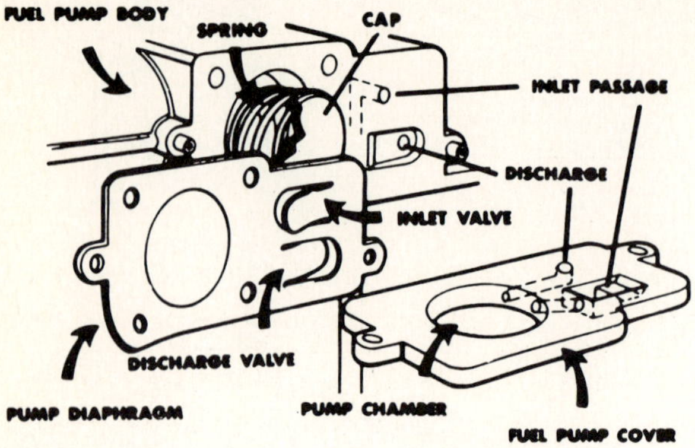

FIGURE 5-23. Pulsa-jet parts breakdown (Courtesy of Briggs & Stratton Corporation).

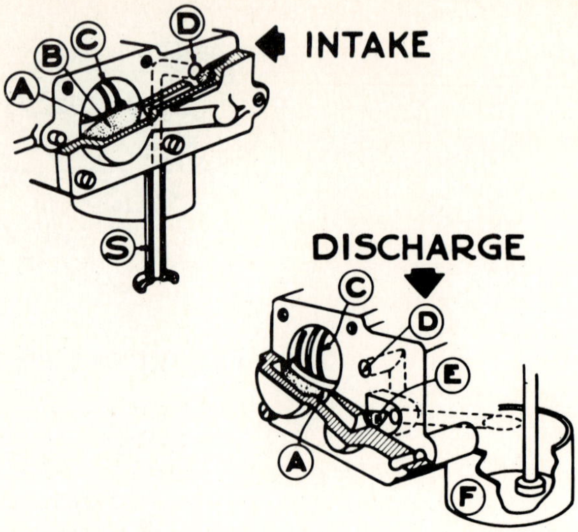

FIGURE 5-24. Pulsa-jet intake and discharge route fuel pump (Courtesy of Briggs & Stratton Corporation).

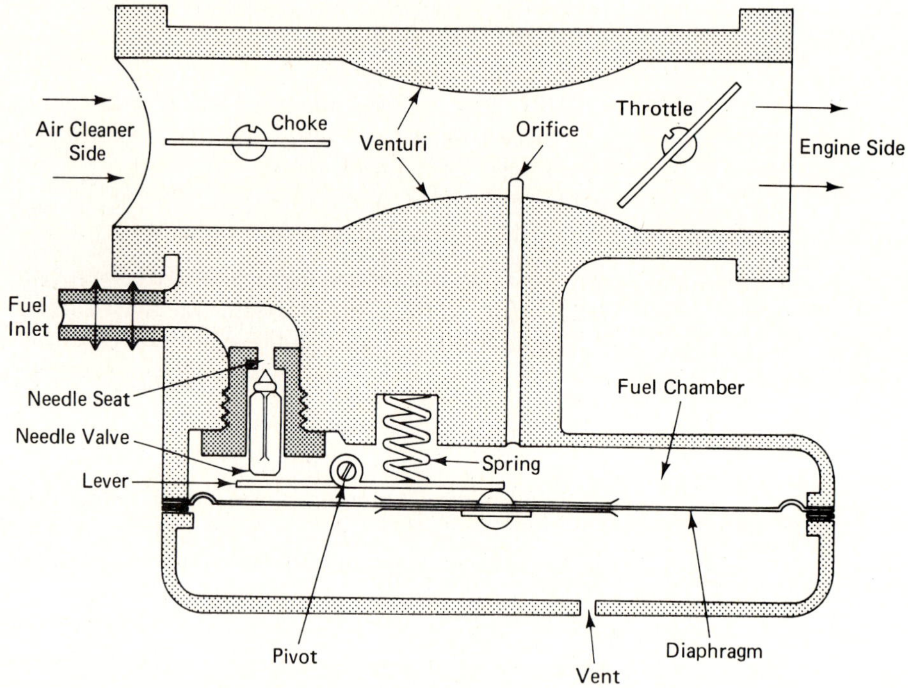

FIGURE 5-25. Basic diaphragm-type carburetor.

separate fuel pump or from a pulse-type pump constructed integrally with the carburetor. Passage into the storage area is controlled by a needle valve. The needle valve's position, in turn, is controlled by a lever arm whose end rests on the center of the diaphragm. As fuel is drawn into the venturi, the diaphragm spring is able to push the diaphragm and needle valve linkage up. This causes the needle valve to open and allow pressurized fuel to enter the storage area. Then, as the fuel enters, it pushes back down on the diaphragm, which causes the needle to shut the entry valve. The principle is the same as a float control system, except that a spring-loaded diaphragm is used instead of a gravity-operated float.

SUCTION-TYPE CARBURETORS

Suction-type carburetors avoid the problem of controlling fuel flow into the temporary storage area by a process of elimination. No temporary storage area is provided. Instead, fuel goes from the tank, which is located beneath the carburetor, directly into the venturi. The pressure differential between the vented tank and the venturi is the sole means of fuel transport.

Naturally, in such a system, special features are provided to control fuel mixing and flow. A spiral is formed in the carburetor throat to give a swirling flow pattern to the air as it passes through the carburetor. This promotes fuel and air mixing, which is especially important during acceleration and sudden load conditions. At those times, the mixture is likely to become excessively lean, causing the engine to stumble and falter.

The amount of fuel delivered at normal operating speeds is controlled by a needle valve and seat assembly. Turning the screw in or out changes the position of the needle, which alters the basic mixture. (See Figures 5-26 and 5-27.)

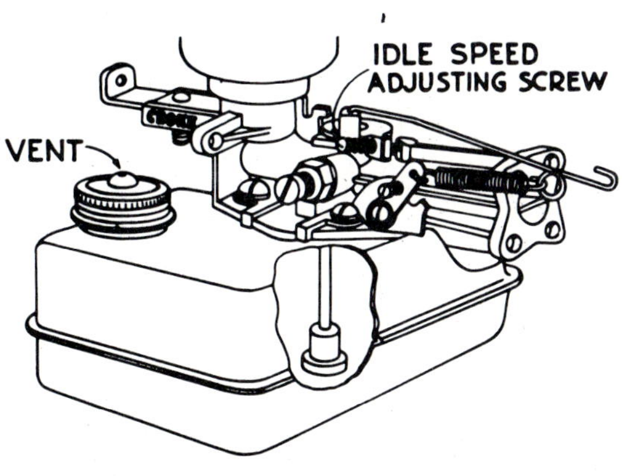

FIGURE 5-26. Vacu-jet (suction-type) carburetor (Courtesy of Briggs & Stratton Corporation).

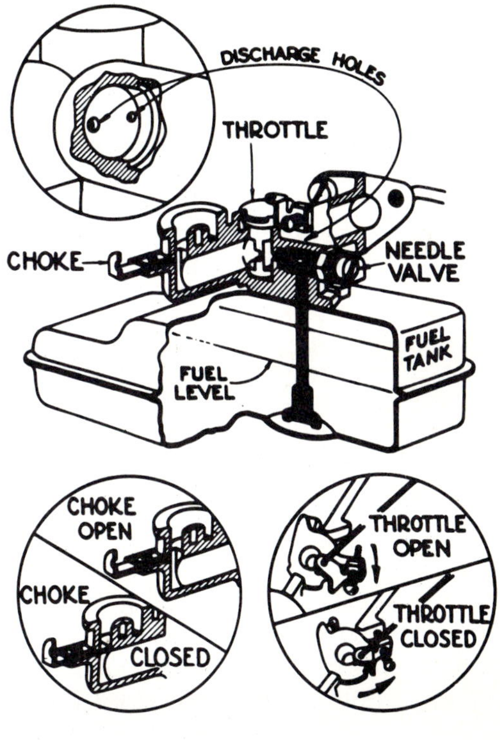

FIGURE 5-27. Cross-section of a vacu-jet (suction-type) carburetor (Courtesy of Briggs & Stratton Corporation).

GOVERNORS

Normally, an engine tends to speed up when the load diminishes and slow down when the load increases. Some small engines use a governor to counteract this tendency. The governor closes the throttle when the engine speeds up and opens when the engine slows down.

Two basic types of governors are found on small gasoline engines: the air vane type and the mechanical or flyweight type. The primary difference between the two is the way in which they sense engine speed.

Air Vane Governors

This type of governor uses the cooling air blown out by the flywheel to determine engine speed. (See Figure 5-28.) The main mechanical component of the system is an air vane connected by a lever assembly to the throttle. The air vane is positioned in the air stream coming from the cooling fan. When the engine speeds up, the velocity of the air from the fan increases, which increases the pressure on the vane. This tends to close the throttle and reduce speed.

Opposed to the force exerted by the air pressure is the governor spring. It is also attached to the throttle and pulls in the opposite direction to open the throttle. The throttle spring is usually connected to a manually operated cable or other adjusting mechanism. By increasing or decreasing tension on the spring, the base throttle setting is changed. The point at which the pull of the spring equals the force of the air vane is called the *governed speed*.

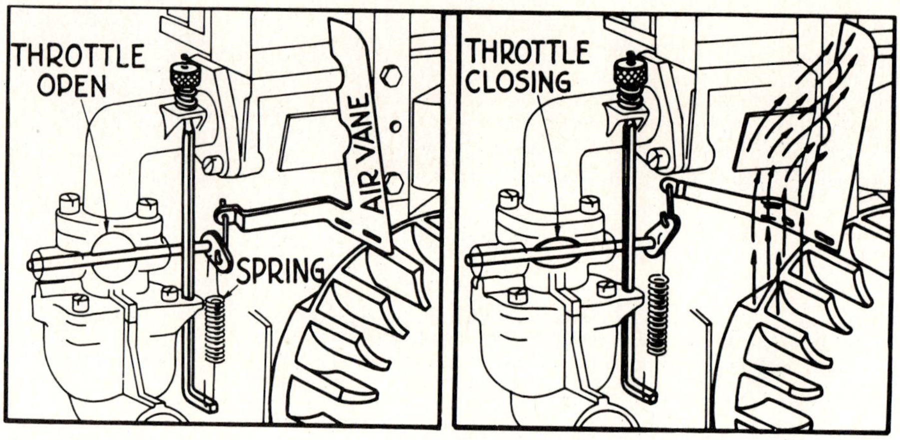

FIGURE 5-28. Pneumatic governor (Courtesy of Briggs & Stratton Corporation).

Mechanical Governors

Mechanical governors use centrifugal force acting on "flyball" weights to gauge engine speed. (See Figure 5-29.)

Centrifugal force, as you may already know, acts on bodies travelling in circular paths, attempting to make them go in straight lines. To see how the principle works, suppose you tie a cord to a rock and swing the rock around your head. Centrifugal force pulls the rock tight against the cord and causes the rock to travel in a flat path around your body. The faster you swing the rock, the greater will be the centrifugal force pulling on the rock.

In a mechanical system, two flyball weights are swung around by the rotation of the crankshaft. When the engine speeds up, the weights swing out; when the engine slows down, the weights drop. The weights are connected to the throttle spring in a somewhat similar manner to the air vane connection described previously. Consequently, when the engine speeds up, the weights swing out and pull on the governor spring. The spring, in return, pulls the throttle in a closed position. Then as the engine slows down, the weights drop, which allows the spring's own tension to pull the throttle toward the open position. The basic throttle setting, as before, is usually determined by a manually operated cable or some other throttle-adjusting device.

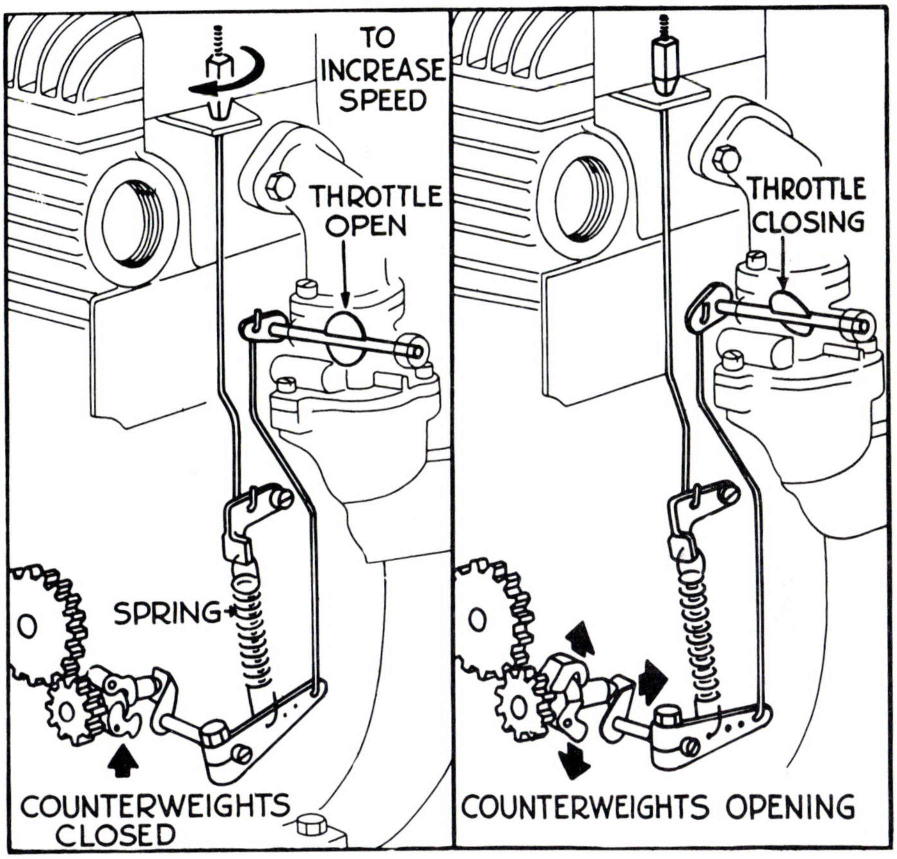

FIGURE 5-29. A mechanical governor (Courtesy of Briggs & Stratton Corporation).

6

ELECTRICITY FUNDAMENTALS

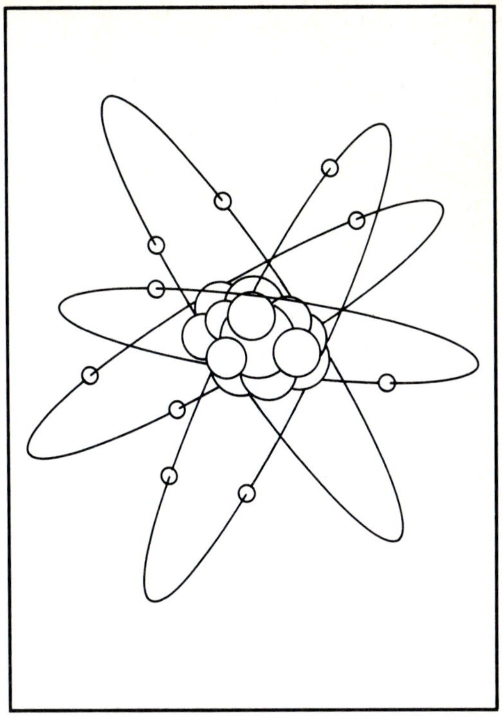

GENERAL

Chapters 7 and 8 examine small engine ignition starting and charging systems. To properly understand these systems you need some background in electricity fundamentals, which is what the present chapter is all about. It will introduce you to:

1. Electrical theory
2. The differences between conductors, nonconductors, and semiconductors
3. Electrical measurements
4. Magnetism
5. The basic operation of some devices that use magnetism
6. Simple electrical circuits
7. Batteries

WHAT IS ELECTRICITY?

First, electricity. We often use the word as if it described a real, physical object. However, that is not exactly what electricity means. How many times have you seen and touched electricity?

When we talk about electricity, we are actually talking above movement, a flow of tiny particles deep within the heart of matter. The visible or detectable signs of electricity—sparks, shocks, readings on test instruments, etc.—are a result of this flow. When the particles are in motion, they have an effect on the materials through which they flow as well as on the surrounding space.

NATURE IN BALANCE

Why do the particles move? It has to do with the universal tendency of conflicting forces to try to stay in balance. The planets remain in their orbits because the centrifugal force of their movement is balanced by the force of gravity from the sun. Water stays in a man-made lake because the weight of the water is balanced by the strength of the dam.

All these are examples of forces in balance. However, what happens if something changes, if, for instance, the dam breaks? Then the forces are no longer in balance. In the case of the broken dam, water rushes downhill, affecting everything in its path, until the lake is emptied and a new balance is achieved.

It is this urge for balance that causes electrical flow. A battery, magneto, or some other device builds up the number of tiny electrical particles in one part of a circuit and reduces the number in another part of the circuit. If the circuit is complete, or, in other words, if there is an uninterrupted path between the out-of-balance parts, current flows. Excess particles flow to the part of the circuit containing fewer particles. The current continues to flow as long as a magneto or battery maintains the imbalance. Like the water flowing downhill from the broken dam, everything that lies in the path of the current is affected.

ATOMIC STRUCTURE

The next question is, what are these particles? If you have been exposed to the subject previously, you've probably already guessed that the particles are electrons. Electrons, along with protons, neutrons, and other particles, make up atoms, which are the basic building blocks of nature. Atoms or pieces of atoms combine in various ways to create all the material in the universe.

The simplest atomic combination is the element hydrogen, which is a very light gas at room temperature. (See Figure 6-1.) It has one proton at

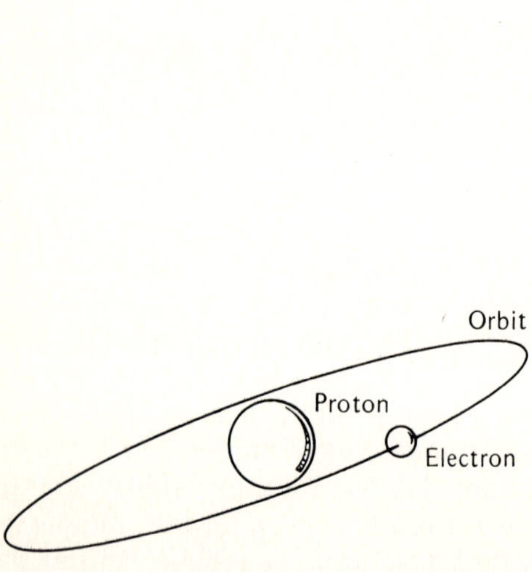

FIGURE 6-1. Hydrogen, the simplest atom.

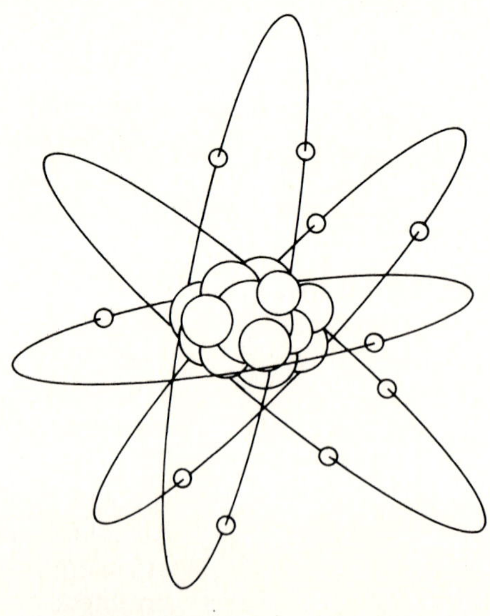

FIGURE 6-2. A more complex atom.

the center of the atom and one electron whirling around the proton, something like a planet whirling around the sun. Hydrogen was one of the first elements created at the beginning of the universe. Other elements, i.e., helium, oxygen, copper, gold, lead – more than one hundred altogether – were formed as additional protons, neutrons, electrons, and other parts were added to the original hydrogen atoms. (See Figure 6-2.)

The atomic parts "stick" together to create atoms because nature strives for balance everywhere, from the very small to the very large. In atoms, the balance we are interested in is between protons and electrons. Protons are said to have a positive charge, like the south end of a magnet. Electrons are said to have a negative charge, like the north end of a magnet. In order for an atom to be balanced electrically, it must have an equal number of electrons and protons.

When an element is created, protons are packed together in the center, or nucleus, of the atom. Electrons are attracted in equal numbers to orbits surrounding the nucleus. Because the electrons have the same charge and tend to repel one another, each orbit can only contain so many electrons, two in the first orbit, six in the second, and so on.

The last or outermost orbit is the most important in our understanding of electricity. The number of electrons in this orbit determines whether an element will be a conductor, nonconductor, or semiconductor of electricity. It is in this orbit that electrical flow takes place.

ELECTRICAL FLOW

To see how electrical flow works, we will take an imaginary trip into the atomic structure of a copper wire. Copper, a conductor of electricity, has 32 electrons. That gives enough electrons to fill all but the last orbit, which has only one electron. Because these single electrons are so far from the nucleus and because they are not bound in a fixed pattern with other electrons, they tend to drift from atom to atom. From our imaginary vantage point, the outer orbit electrons might resemble a swarm of fireflies, wandering here and there in response to random influences.

As long as the wire is not hooked into a live electrical circuit, the pattern of electron movement remains random. An atom in one place loses an electron. Another electron, from somewhere else, takes its place, attracted by the extra positive charge exerted by the atom that has lost its electron. Since the movements are random, the sum total of all the electrical charges on all the atoms remains balanced. (See Figure 6-3.)

However, what happens if the ends of the wire are attached to the positive and negative terminals of a battery? The negative pole contains an excess of electrons, and the positive pole has fewer electrons than it needs to be balanced. The excess electrons at the negative pole repel the electrons in the wire, causing them to flow toward the other end. At the same time, the electrons at the other end of the wire are attracted to the empty orbits at the battery's positive pole. Almost simultaneously with that, the excess electrons from the battery's positive pole rush into the wire to fill the newly vacated orbits. (See Figure 6-4.)

It's like a game of tag played at the speed of light, where nobody (or no electron) ever catches anybody. The chemical activity inside the battery constantly creates an imbalance between the two poles. So long as the battery remains active and the circuit is complete, the electrons will chase

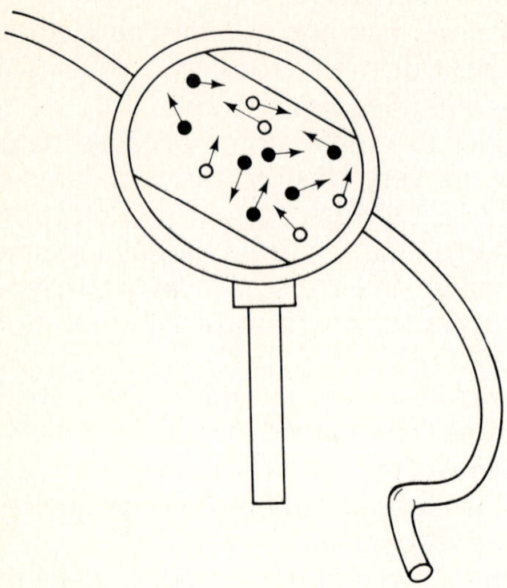

FIGURE 6-3. Random electron movement.

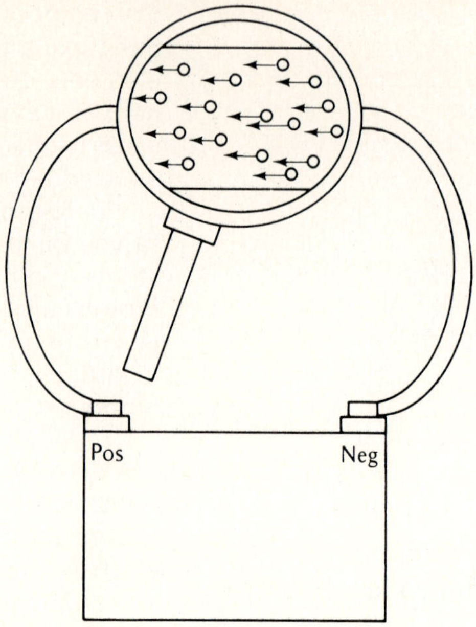

FIGURE 6-4. Directed electron movement.

each other from empty orbit to empty orbit, through the battery and through the wire. If the battery weakens, the current slows down. If the wire is cut, then the flow stops altogether. However, the battery still pulls electrons away from the end of the wire connected to the positive pole and builds up electrons at the end connected to the negative pole.

Electrical flow takes place wherever there is an overabundance of electrons in one place, an underabundance in another place, and an uninterrupted path between the two. The potential for electrical flow exists whenever there is an imbalance but no complete path for current to flow. For instance, even though no current flows when the ignition switch in a car is off, the potential for current flow still exists because there is an electrical imbalance on either side of the switch.

Current flow can be temporary or sustained. The sudden discharge of electricity represented by a lightning bolt is obviously temporary. An electrical imbalance is created between a cloud and the ground. When the potential becomes great enough, the excess electrons are able to create a path through the atoms in the air.

Most sustained current flow is the result of some man-made device: a battery magneto, alternator, generator, etc. They convert either chemical or mechanical energy into electrical potential. That is, they create regions of electrical imbalance, which, when connected to a complete circuit, will cause electrical flow.

CONDUCTORS, NONCONDUCTORS, AND SEMICONDUCTORS

The number of electrons in the outer orbit of an atom determines how readily substances containing the atom will allow electrical flow.

Conductors

Substances containing atoms with one to three outer orbit electrons are usually considered to be good conductors. The electrons in their outer orbits are loosely held and easily put into motion. Copper, gold, iron, silver, and water are examples.

Nonconductors

Materials containing atoms with five or more outer electrons are poor conductors under normal circumstances. Their electrons are tightly held in a pattern with other electrons and are not easily dislodged. Nonconducting materials like rubber, glass, and many plastics are used as insulators to channel and direct electrical flow—making sure it goes only where it is supposed to go.

Semiconductors

Semiconductors are very stable elements. Their atoms, containing four outer electrons each, join together in rigid patterns that are not easily disturbed and that do not readily support electron flow. However, this holds true only so long as the element remains in its pure state. If small amounts of impurities are added, the semiconductor becomes a conductor. Depending on the kind of impurity added, current flow will depend primarily on the presence of positively charged "holes" (actually vacant spaces in outer orbits) or on free electrons floating through the semiconductor. (See Figure 6-5.)

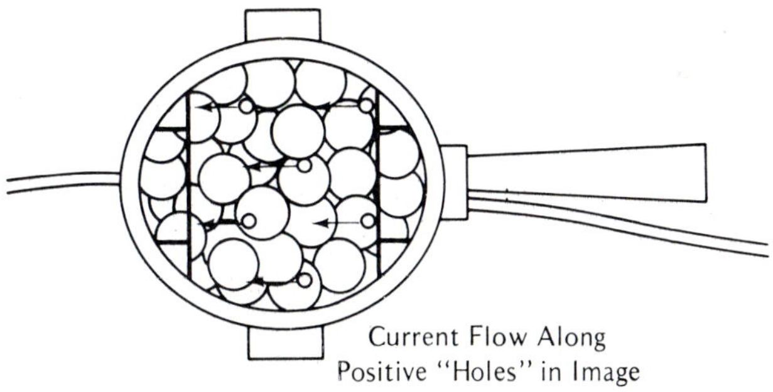

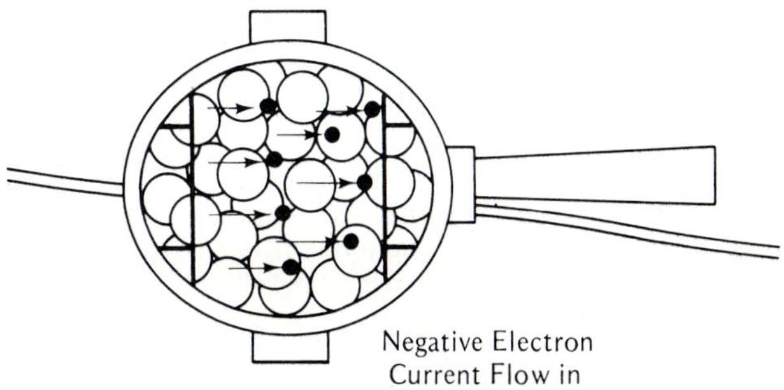

FIGURE 6-5. Current flow through two types of semiconductors.

In small engine applications, semiconductors are used primarily to make diodes and transistors. Diodes are formed when two different kinds of "impure" semiconductor chips are sandwiched together. Diodes allow current flow in only one direction. They are like one-way current valves (see Figure 6-6). Transistors are formed by sandwiching three semiconductors together. The two outside layers are made from one type of semiconductor and the middle layer from the other type.

The middle layer is often connected to a low-power control circuit. The outer layers are connected to a higher-powered main circuit. Owing to the way the transistor functions, low-level fluctuations in the control circuit will cause corresponding (but magnified) changes in the main circuit, including shutting it off if the control circuit is shut off. In this way, the transistor acts as a kind of amplifier. (See Figure 6-7.)

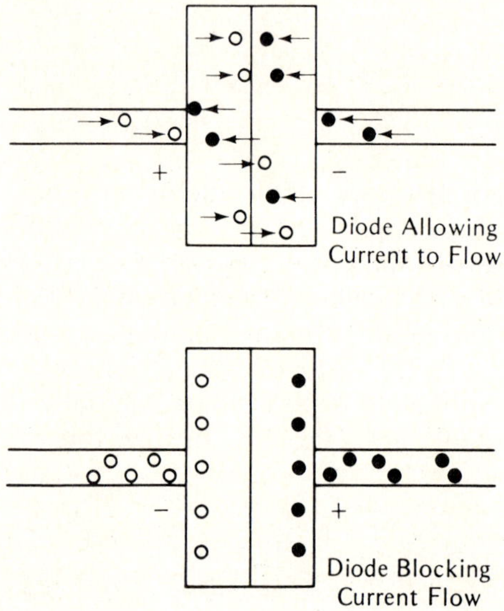

FIGURE 6-6. Diodes as one-way valves.

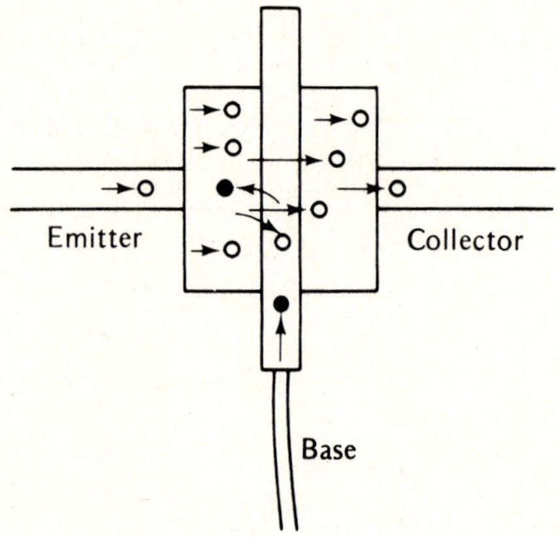

FIGURE 6-7. Transistors.

ELECTRICAL MEASUREMENT

We have seen now what electrical flow is and why and how it works. The last item to be considered before going on to magnetism is electrical measurement.

Volts

Voltage is a measure of electrical pressure. It is an indication of the force exerted on each electron in a circuit by a source of electrical potential. As the following example shows, it is not a measure of the total electrical force felt by all the electrons in a circuit.

To see what this means, imagine that you have two 12-volt batteries, a large one and a small one. Both have the potential for exerting 12 volts of force or pressure on each electron in a circuit. In other words, the electrical imbalance existing at the positive and negative poles of both batteries will have the same disruptive effect on a given electron in a given atom. However, the larger battery has the potential for disturbing more electrons and more atoms.

Amperes

Amperage is a measure of total electrical flow. It is an indication of the number of electrons flowing past a given point in a circuit. Amperage (or the number of electrons in motion) depends on electrical pressure (voltage) and on electrical resistance (ohms, discussed next).

Ohms

An ohm is a unit of electrical resistance. It is used to express the varying ability of different materials to support current flow. Some factors affecting resistance are: (1) the size of the conductor—just as a larger water pipe will let more water flow than a small pipe, a large conductor will let more electrons flow; (2) the length of the conductor—the longer the conductor, the greater the resistance; and (3) the nature of the conductor—as we have already seen, some materials support current flow better than others.

OHM'S LAW

As you might imagine, ohms, amperes, and volts are related quantities. If one quantity changes, at least one of the others must also change. For instance, in a given circuit:

1. If voltage changes (goes up or down) and the ohms of resistance stay the same, then the amperage must also change (go up or down).
2. If the ohms go up (as a result of a frayed, or otherwise damaged wire) and the voltage at the source doesn't change, the amperage must go down. Not as many electrons can flow.
3. Conversely, if the ohms of resistance drop, the amperage must increase.

These relations are expressed in a statement called Ohm's law. That law and associated formulas are given in Figure 6-8.

Ohm's Law:

An electrical pressure of 1 volt is required for 1 ampere of current to flow past 1 OHM of resistance.

Stated as a formula:

$$E = I \times R$$

or

$$I = \frac{R}{E}$$

or

$$R = \frac{E}{I}$$

E = Voltage (or Electromotive Force)

R = Resistance

I = OHMS (or Impedance)

FIGURE 6-8. Ohm's Law.

WHAT IS MAGNETISM?

Magnetism is a mysterious force operating equally well through air, empty space, or solid matter. No one knows exactly how it works. Useful theories have been devised that help predict how magnetism behaves and how it can be put to practical application. We will examine some of these theories because they are needed to help understand small engine electrical operation.

Like electricity, magnetism can be neither felt, seen, nor touched. It is more a description of an effect than of a thing.

WHAT IS A MAGNET?

A magnet itself is easier to describe than magnetism. Magnets are objects with the power to attract or repel other magnets and to attract iron and certain materials made from iron. Magnets are directional; that is, they have distinct ends, or poles. If a straight or bar magnet is allowed to hang freely in a horizontal position, one end (called the north pole) will always point in a northerly direction and the other end (called the south pole) will always point toward the south. No matter how a magnet is cut, shaped, or altered, it (or its pieces) will have two magnetically different poles.

THEORY OF MAGNETIC OPERATION

Scientists believe that magnetic properties originate at the atomic level. This is a slightly altered version of the explanation most often given.

Every atom in the universe is thought to spin on its axis, like the earth spins on its axis, or a toy top spins about its handle. The spinning action (along with electron movement) causes the atoms to be surrounded by lines of magnetic force. No one knows exactly what these lines of force are, or if they even exist in the usual physical sense. However, for our purposes, we can visualize them as being the paths taken by imaginary particles flying around and through atoms. The particles travel out one end of an atom's axis (its north pole), circle around the atom, reenter at its south pole, and then go through the atom to start the trip again. (See Figure 6-9.)

In most materials, the atomic poles point in different directions. The lines of force surrounding adjacent atoms don't line up; the imaginary particles flying around neighboring atoms bump into one another as often as they travel along parallel paths.

However, something happens in magnetic materials. A substantial number of atomic poles point in the same direction. This means that a substantial number of the imaginary particles travel in the same direction. Instead of expending their imaginary energy in headlong collisions, the nonexistent particles fly along parallel paths. They join forces to fly from atom to atom, building up enough momentum to actually escape, for a while, the confines of the magnetic material. (See Figure 6-10.)

The action duplicates on a large scale what takes place on a small scale within atoms. Imaginary particles fly along parallel paths within a magnet, building up enough power and energy to leave the magnet at its north pole. Then the particles circle back to reenter the magnet at its south pole and start the trip again. The paths taken by the particles represent lines

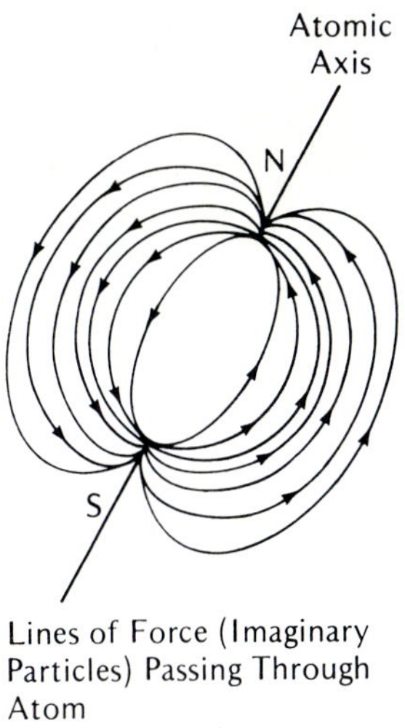

FIGURE 6-9. Magnetic force lines passing through a single atom.

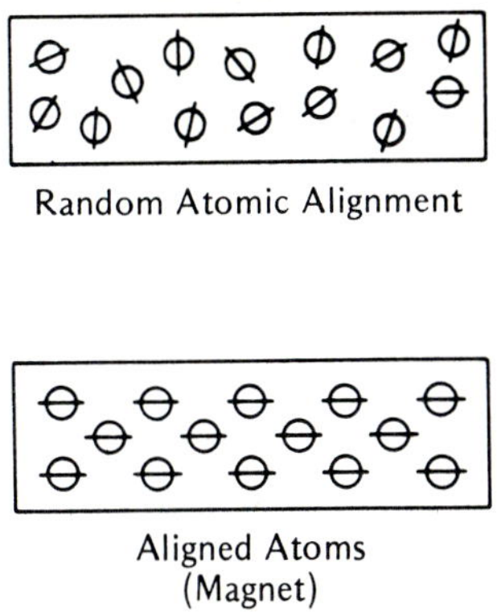

FIGURE 6-10. Aligned *vs.* unaligned atoms.

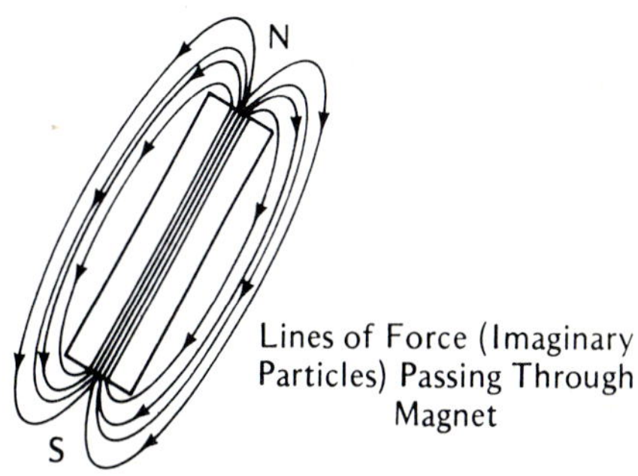

FIGURE 6-11. Lines of force passing through a bar magnet.

of magnetic force—which extend in three dimensions on all sides of the magnet. (See Figure 6-11.)

The flight of the imaginary particles helps to explain how magnetic materials attract and repel each other. When the opposite poles of two magnets are placed together, the imaginary particles travel in the same direction, just as they do within the atomic structure of the two magnets. Consequently, the particles join forces, pulling the two magnets together. In effect, this creates a single magnet with double the power of the individual magnets.

However, if the like poles of two magnets are put in close proximity, the particles move in opposite directions. As a result, they bump together to push the magnets apart. (See Figure 6-12.)

Magnets are attracted to iron objects when the lines of force from the magnet penetrate the atomic structure of the iron material. The iron atoms tend to line up with the lines of force, causing the imaginary particles to fly along in the same paths. As the particles join forces, they tend to pull the two objects together.

These magnetic flight paths (whatever they represent) can actually be seen by sprinkling iron filings on a piece of paper that has been placed over a magnet. Tapping the paper lightly loosens the filings so that they can follow their natural attraction to the lines of force. The pattern of filings represents the shape of that part of the three-dimensional magnetic field sliced through by the paper. Figure 6-13 pictures the magnetic fields surrounding horseshoe and the bar magnets.

KINDS OF MAGNETS

There are three basic kinds of magnets: permanent, temporary, and electromagnetic.

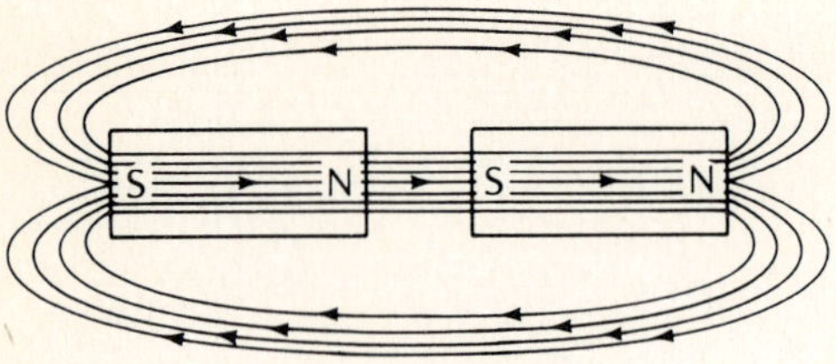

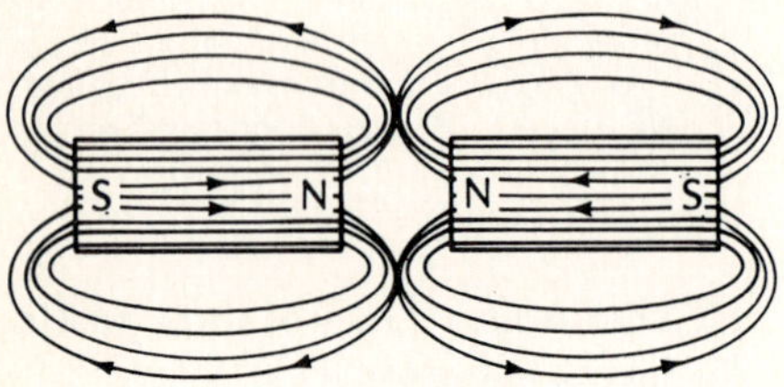

FIGURE 6-12. Attraction and repulsion between two magnets.

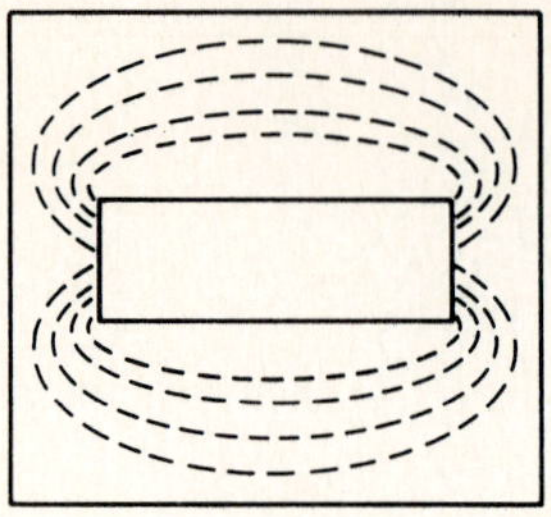

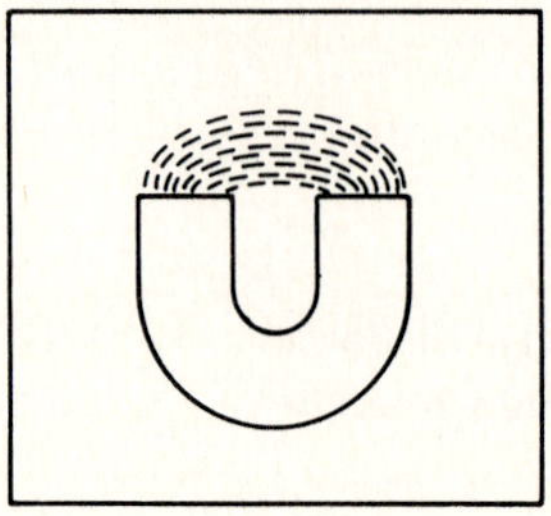

FIGURE 6-13. Iron filings around horseshoe and bar magnets.

Permanent Magnets

Permanent magnets are often made from the mineral magnetite, which is a naturally occurring magnet. Its atoms are sufficiently aligned to produce a coherent magnetic field. Commonly called "lodestone," magnetite was discovered hundreds of years ago. Early mariners used it as a crude compass to help chart their course when travelling out of sight of land.

Temporary Magnets

Temporary magnets are created by placing certain iron or iron-based materials in the presence of strong magnetic fields. The lines of force from the magnetic field "pull" the iron atoms into alignment. Depending on the strength of the magnetic field and on the nature of the material, the atoms may remain in alignment after the magnetic field is removed. However, as the name implies, these kinds of magnets may not last very long. Any sort of shock can disturb the fragile atomic alignment; this causes the atoms to become jumbled again and no longer capable of producing a coherent magnetic field.

Electromagnets

Electromagnets depend on a peculiar property of electricity: whenever current flows through a conductor, the atoms in the conductor line up sufficiently to produce a coherent magnetic field.

If the conductor is a wire, the lines of force form concentric rings around the wire. In other words, the imaginary particles fly in circular paths around the wire. The direction in which the nonexistent particles fly, clockwise or counterclockwise, depends on the direction in which the current is flowing. (See Figure 6-14.)

Electromagnets are created when current-carrying wires are formed into a coil. This is how it works. Current flows in the same direction in adjacent loops of the coil. Consequently, the imaginary particles fly in the same direction in adjoining loops. As we have seen before, whenever these particles fly in the same direction, they join forces. The lines of force merge to circle all the loops, travelling around the outside of the coil, going back through the inside of the coil, then going around the outside again. (See Figure 6-15.)

An electromagnet is like a permanent magnet in many ways. It has a north pole (where the particles come out of the coil) and a south pole (where the particles return). An electromagnet can attract or repel other magnets. It can also attract iron.

However, there are still some important differences between the two. First of all, electromagnets can be turned on or off by turning the

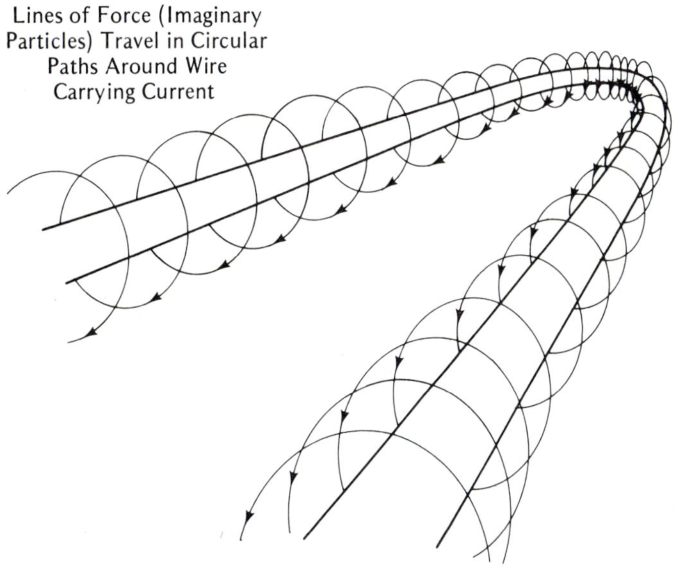

FIGURE 6-14. Force lines going around wire.

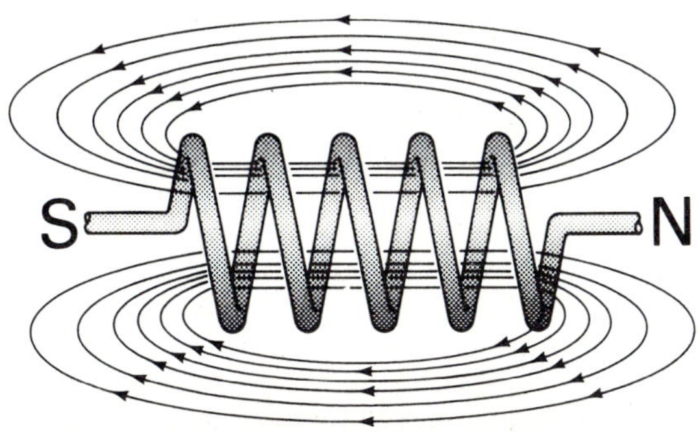

FIGURE 6-15. Wire formed in an electromagnetic coil.

current on or off. The polarity of the electromagnet (which end is north and which is south) can be reversed by reversing the current flow. Also, the strength of the magnet can be altered by varying the current flowing through the coil or by increasing the number of loops in the coil.

APPLICATIONS OF MAGNETISM

We would not study magnetism if it did not have some practical use. The following paragraphs briefly introduce the operation of solenoids, relays, electric motors, and induced current flow devices.

Solenoids

Solenoids are electromagnetic coils used in mechanical actuating devices, attracting or repelling parts of a mechanical linkage. A solenoid coil is usually hollow. Often called a "sucking coil," it has the power to attract objects inside the coil itself—or to repel objects from the coil. (See Figure 6-16.)

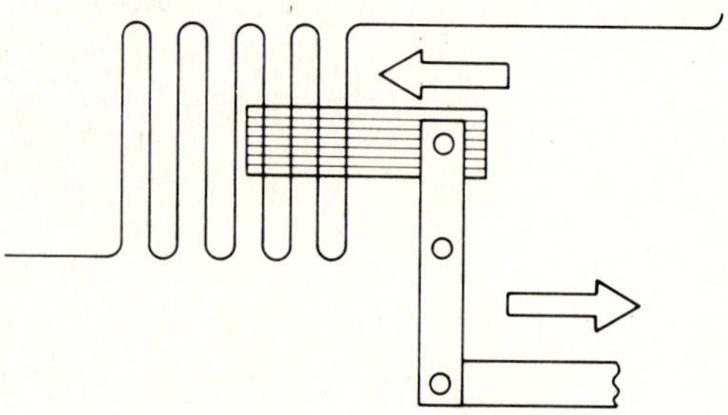

Solenoid . . . Coil Used to Actuate Mechanical Linkage

FIGURE 6-16. A solenoid.

Relays

A relay is an electromagnetic coil used (most often) in an electrical switch. Usually the relay coil is wrapped around an iron core. That is because metals with a high degree of magnetic "permeability" possess the power to concentrate lines of magnetic force, thereby increasing the strength of the magnet in the region of the core. It is as if the imaginary particles prefer to pass through certain materials. Consequently, though the total number of force lines is not increased, the concentration in certain regions is increased. (See Figure 6-17.)

Electric Motors

Electric motors contain two sets of magnets, the field and the rotor. The field may be an electromagnet or a permanent magnet. The rotor is always an electromagnet.

Figure 6-18 illustrates a very simple electric motor. The field is a

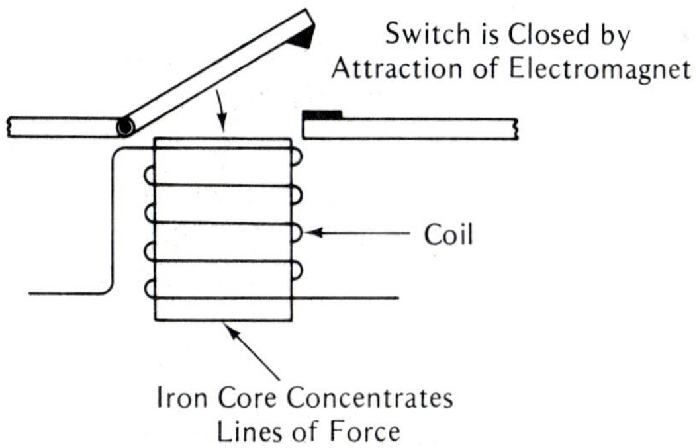

FIGURE 6-17. A relay.

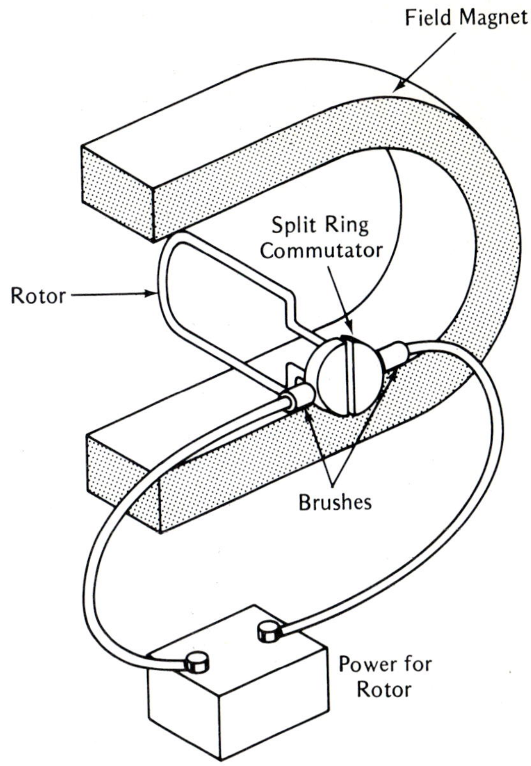

FIGURE 6-18. An electric motor.

horseshoe magnet, and the rotor is a single loop of wire placed between the ends of the horseshoe magnet. The rotor is connected by a switch (called a split ring commutator) to the power source.

In terms of the imaginary particles, this is how the motor works: Particles travel from the north to the south pole of the field magnet. At the same time, particles circle the rotor loop, going around one way as the current flows down one side of the loop, and revolving the other way as the current flows back.

The end view of the motor shown in Figure 6-19 shows how the particles interact to turn the rotor. The top part of the rotor is pulled to the left

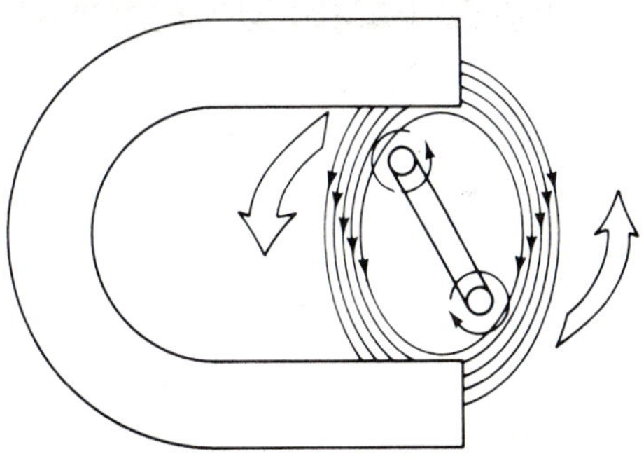

Force Lines Push and
Pull Rotor

FIGURE 6-19. An end view of a motor showing the reaction between field and rotor force lines.

because the particles from the field and the rotor move in the same direction on the left side of the loop and in the opposite direction on the right side of the loop. The particles on the left trying to join the particles on the right push apart—which pushes the top to the right.

A similar action takes place at the bottom of the loop. However, since the current (and hence the particles) move in the opposite direction, the pushing/pulling is to the right.

These forces move the rotor halfway around, as noted in Figure 6-20. Momentum then takes the rotor on a little further. However, unless something happens, the forces acting on the rotor will cancel out at this point. They will work in opposite directions, and the rotor will stop.

That is where the split ring commutator comes into play. As noted in Figure 6-21, it reverses the current flow through the loop. The two halves of the commutator swap connections with the power source. What was once the negative connection to rotor becomes the positive lead and vice versa. As a result, the particles or lines of force circle the loop in opposite directions. Therefore, the forces acting on the rotor are reversed, causing it to continue on around. This cycle is repeated once every revolution of the rotor.

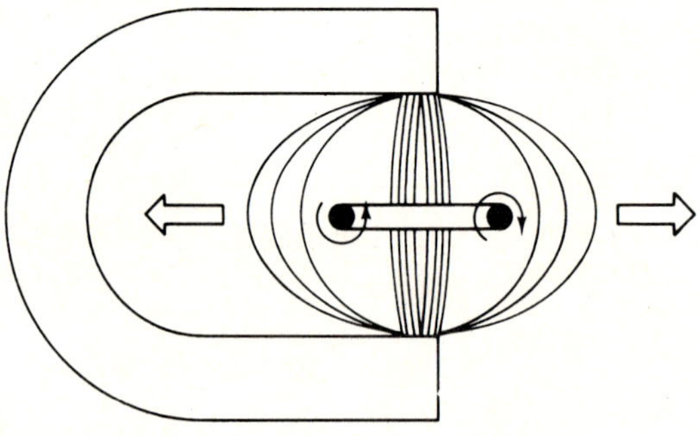

FIGURE 6-20. A rotor after turning halfway around.

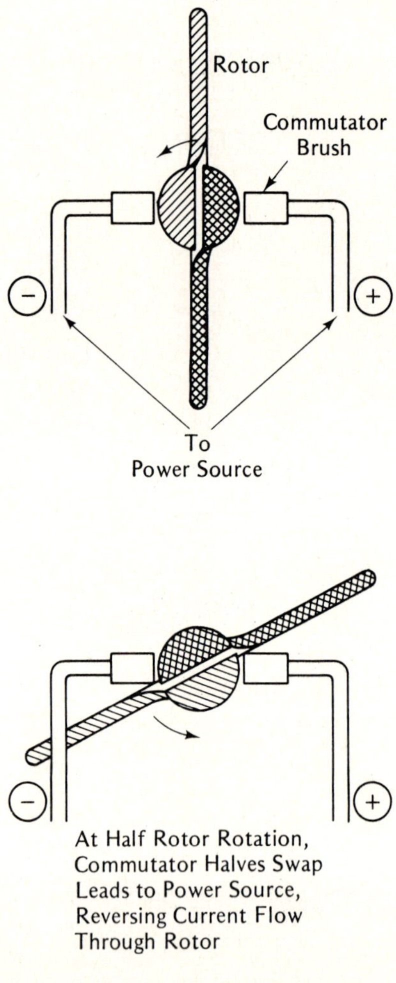

FIGURE 6-21. A commutator reversing current flow.

Electric Generators (Induced Current Flow)

Theoretically at least, most electric motors can be turned into electric generators. (Some small-engine starter motors double as generators.) It is possible because of a peculiar property of conductors and magnetism called *induced current flow.*

Whenever a conductor and magnetic lines of force move in relation to one another, current flow is generated (or induced) in the conductor. For instance, if one side of a loop of wire is passed between the ends of a horseshoe magnet, current will flow through the wire. The same thing happens if the wire is held stationary and the ends of the magnet move. (See Figure 6-22.)

No one knows exactly how current flow is induced. The explanations that do exist are very complicated. For our purposes, we can simply say that the imaginary particles knock electrons loose from the outer orbits of atoms.

The direction in which the electrons travel depends on the relative motion between the conductors and the lines of force. If the imaginary particles cut through the atomic structure of a conductor from one direction,

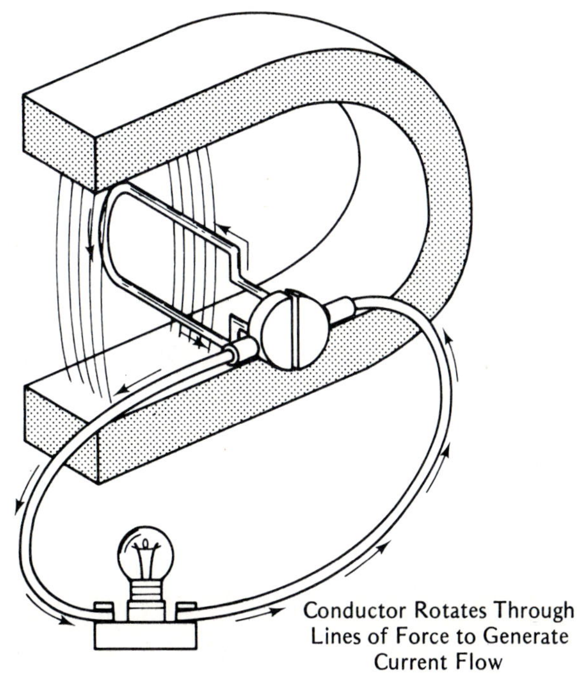

FIGURE 6-22. The motor shown in Figure 6-21 converted to a generator.

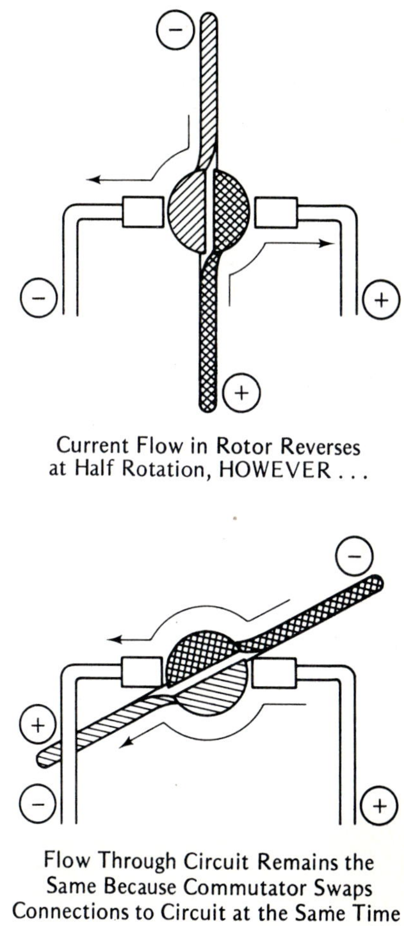

FIGURE 6-23. The action of a commutator converting AC to DC.

the electrons will be knocked one way. If the particles strike in the other direction, the electrons will move that way.

As a result, the simple electric motor described before can be converted into a generator by turning the rotor between the ends of the horseshoe magnet. As the two sides of the loop cut across the magnetic field, current is induced. It flows through the ends of the loop, through the two halves of the commutator, and on into the attached circuit.

However, something happens when the rotor reaches the midpoint of its travel, as noted in Figure 6-23. The two sides of the rotor loop start to cut back through the lines of force in an opposite direction. Consequently, the current will flow in an opposite direction.

In our simple motor/generator, the commutator can be used to keep the current flowing in the same direction. As the current reverses in the rotor, the two parts of the commutator swap sides of the circuit. The commutator half that was feeding one half of the circuit now feeds the other half and vice versa. Therefore, as the current flow in the rotor is reversed, the leads from the rotor to the circuit are also reversed. Consequently, the flow throughout the circuit remains the same.

That is how mechanical force is converted into electrical potential. This potential can be used to operate other devices in the circuit or to recharge a battery.

Other Applications of Induced Current Flow

Electrical generators like those just described are used most often in "larger" small engines. Generators supply charging current to batteries, which in turn operate the electrical devices associated with the engine. "Smaller" small engines employ induced current flow primarily in the operation of magnetos and ignition coil transformers. A magneto is a simple, permanent magnet generator. Unlike generators used to charge batteries, it has no commutator for converting AC to DC current. Because it employs permanent magnets instead of electromagnets to supply magnetic lines of force (unlike most charger generators), its output is not ajdustable. (See Figure 6-24.)

An ignition coil transformer, in its simplest form, is nothing more than two adjacent coils of wire. One coil is supplied with current. Every time the current flow is interrupted, the lines of force surrounding the coil fall across the other coil. The relative movement between the lines of force and the coil induce voltage in its windings. By varying the turns of wire that make up the two coils, the voltage output of the second coil can be made greater than the first. (See Figure 6-25.)

Magnetos and coils will be discussed more completely in the next chapter.

ELECTRICAL CIRCUITS

Whenever a conductor moves in relation to magnetic lines of force or is connected to a source of electrical potential, electrons flow through the conductor. The endless path along which the electrons flow is called an electrical circuit. The following paragraphs examine the basics of electrical circuitry.

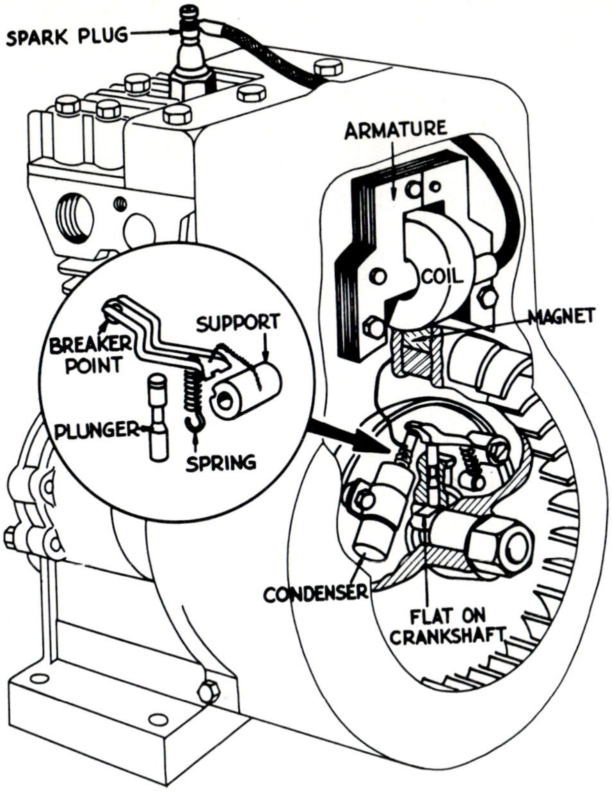

FIGURE 6-24. Magneto components.

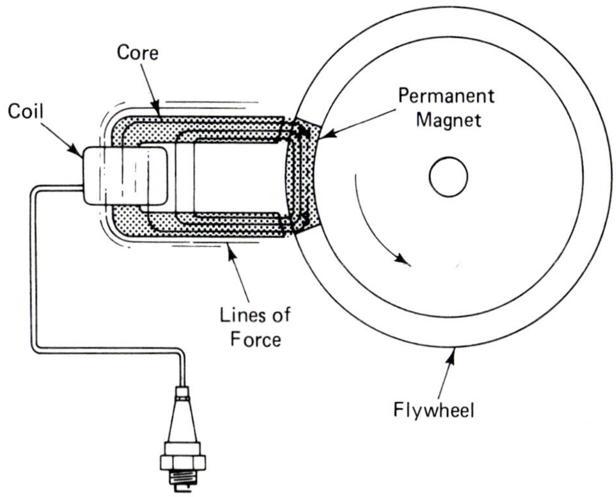

FIGURE 6-25. Magneto operation.

SERIES AND PARALLEL CIRCUITS

All circuits fall into one (or a combination) of two categories: series and parallel.

A series circuit is like a one-way road with no side or connecting roads. All electrons in motion in a series circuit must travel past every part and through every device in the circuit. All points of resistance must be overcome by every electron; any break in the circuit will stop current flow in the entire circuit. Figure 6-26 pictures a simple series circuit.

Parallel circuits are akin to a branching network of roads. Electrons may move along any one or all of the branches. A break in one part of the

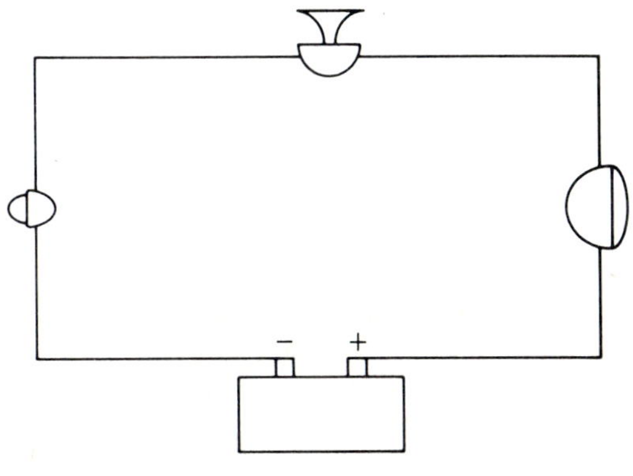

FIGURE 6-26. A simple series circuit.

circuit does not necessarily interrupt the current flow to the other parts. By the same token, changes in resistance (either high or low) in one branch do not affect current flow to the other branches. Figure 6-27 pictures a simple parallel circuit.

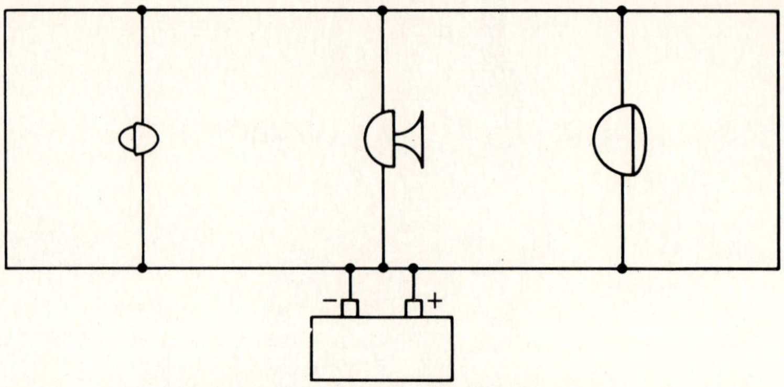

FIGURE 6-27. A simple parallel circuit.

ELECTRICAL SCHEMATICS

The schematics, or wiring diagrams, used to picture electrical circuits are like road maps showing the main features, junctions, and points of interest along electrical pathways. Just as a road map lets a driver plan a trip between two geographical points, a schematic lets a mechanic chart a diagnostic course between the elements in an electrical circuit.

The items pictured on electrical diagrams include switches, batteries, alternators, solenoids, motors—virtually anything that might be found in a circuit. In some cases an item might be pictured literally.

However, in most cases, circuit components are shown symbolically by characters or figures. Figure 6-28 pictures some common electrical symbols.

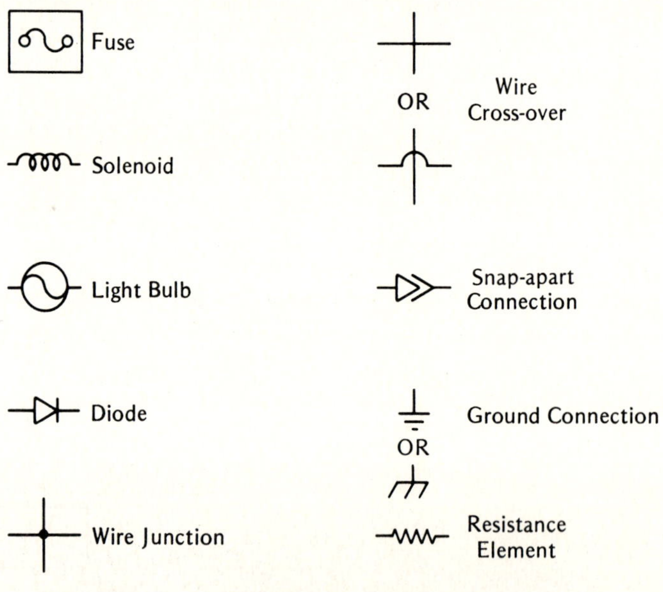

FIGURE 6-28. Some common electrical symbols.

Electrical schematics often appear very complicated. One secret to using them is to concentrate on the problem at hand – since the only reason to go to a wiring diagram in the first place is to diagnose an electrical malfunction. In other words, after locating the likely trouble spots on the schematic and finding the segments of the circuit that connect these elements, ignore everything else.

CIRCUIT FEATURES AND CONDITIONS

However, even before applying this rule, you need to know certain facts about electrical circuits. You wouldn't attempt to use a road map without knowing the difference between various kinds of roads, intersections, etc. You also can't expect to use a wiring diagram without knowing something about various circuit features and conditions.

Ground Return Circuits

In the simple circuits illustrated in the preceding, the current flows along wires to and from the power source. However, in small engines, the conductive parts of the engines (as well as any framework to which the engine is attached) are used for half of the circuitry. The negative terminal of the battery or generating device is connected to the engine. Electrons from the battery or generating device flow into any wire or component connected to the engine. Current flows back to the positive terminal along wires and cables. (See Figure 6-29.)

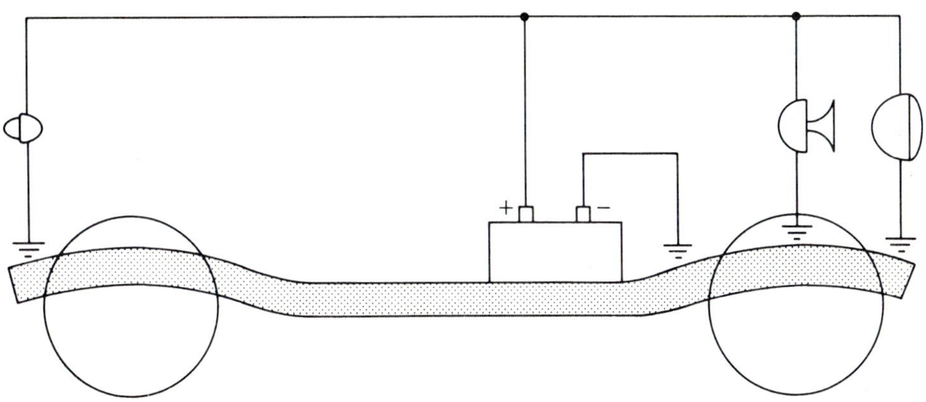

FIGURE 6-29. Ground returns in a small engine circuit.

Open Versus Closed Circuits

Current flows only if there is a closed path between the ground point and the positive terminal of the battery or generating device. So long as there is an intentional break (like an open switch) or an unintentional break (like a severed wire), electrons will not flow. Only the potential for flow will exist – represented on the negative side of the opening by an excess of electrons and on the positive side by a reduction of electrons. (See Figure 6-30.)

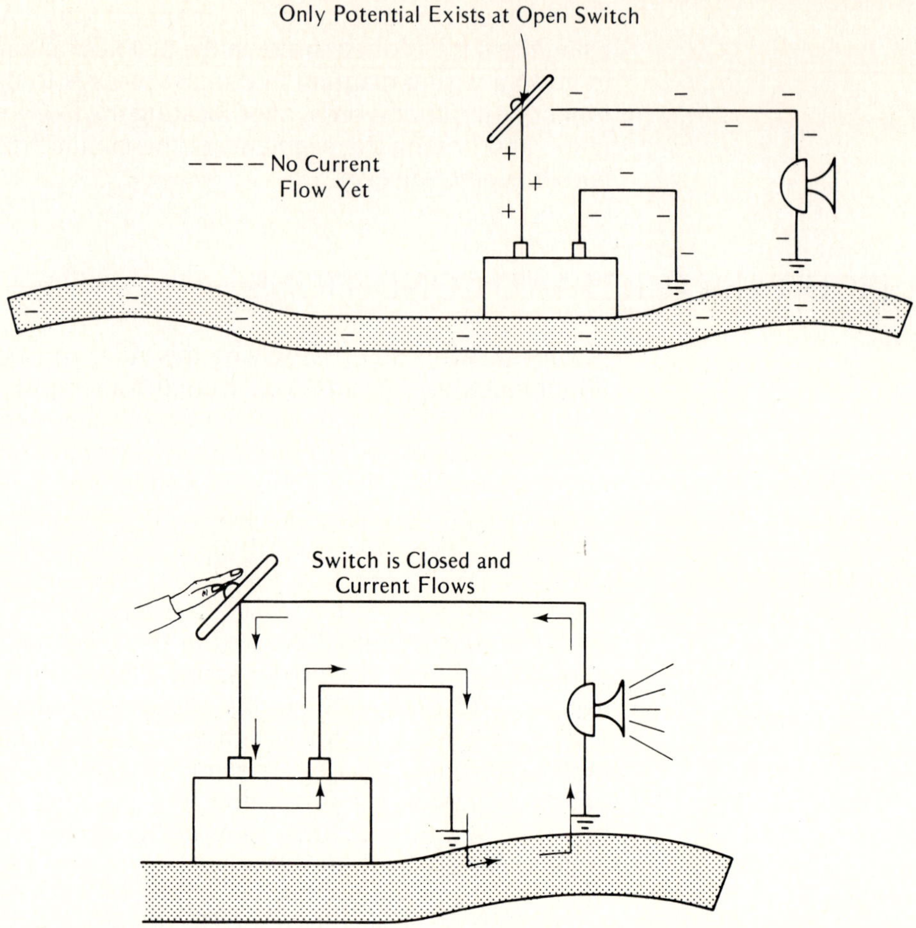

FIGURE 6-30. Electrical potential *vs.* electrical flow at open and closed switches.

CIRCUIT ABNORMALITIES

So far, we've examined various aspects of normal circuits. Next, we will look at some abnormal or problem conditions.

Short Circuits

A short circuit occurs when two wires come into accidental contact. A short represents a junction where none was intended. Because current flows to areas of least resistance, it is possible that all or a substantial portion of the current flowing in both of the shorted wires will be shunted through the one with the lowest resistance. As a result of the added current load, that wire may overheat, possibly enough to cause a fire. By the same token, the wire with reduced current flow will not be able to supply the devices to which it is connected.

Grounded Circuit

A grounded circuit is like a short circuit with one important exception. Instead of two wires coming into accidental contact, a wire touches ground where it shouldn't (e.g., the wire touches part of the engine).

Open Circuit

An open circuit refers to an unintentional break due to a damaged connection, a severed wire, etc.

High Resistance Circuit

Excessive resistance occurs when a wire or connection is damaged or corroded enough to restrict current flow but not block it altogether.

BATTERIES

As noted before, most "small" small engines don't use batteries, relying instead on magneto generating systems. However, many "larger" small engines do use batteries, plus a bona fide generator to recharge the battery when it runs down. The remaining paragraphs in this chapter will introduce battery operation.

In its simplest form, a battery contains: (1) a pair of dissimilar metal plates, (2) positive and negative terminals attached to these plates, (3) a chemically active liquid called electrolyte in which the plates are immersed, and (4) a container for the plates and the electrolyte. (See Figure 6-31.)

Electrons are moved because of the chemical reaction between the electrolyte and the two kinds of plates. Very roughly speaking, the electrolyte allows electrons to move away from one kind of plate and build up on another. As a result, one kind of plate and its attached terminal will have more than the normal number of electrons. The other plate and its terminal will have fewer electrons. In the plate with fewer electrons, the positive protons will predominate, giving that plate and its terminal a positive charge. The plate and terminal with more electrons will have a negative charge.

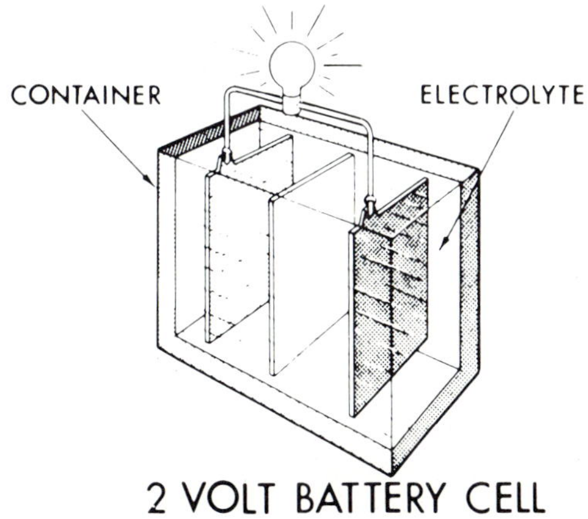

FIGURE 6-31. A wet battery is a device that is able to convert chemical energy to electrical energy. The chemical action can be reversed to recharge the battery (Courtesy of Ford Motor Company of Canada, Ltd.).

These same basic features can be expanded in a number of ways to create a more complex battery. For instance, extra pairs of positive and negative plates can be added, in a kind of alternating "sandwich" (so long as the layers don't touch). These extra plates will not increase the voltage or "push" given to any single electron released by any particular plate. However, the capacity of the battery is increased by increasing the total number of electrons put into motion.

The simplified battery can also be combined with other batteries. Its positive and negative terminals can be joined to the opposite terminals of a second battery, and that battery's terminals to the opposite terminals of a third battery, and so on. Joining batteries this way (positive to negative to positive to negative to positive, and so on) will add to the total electrical "push" or voltage. For instance, if six two-volt batteries are joined together, the total pressure will be (6×2) 12 volts. (See Figure 6-29.)

7

IGNITION SYSTEMS

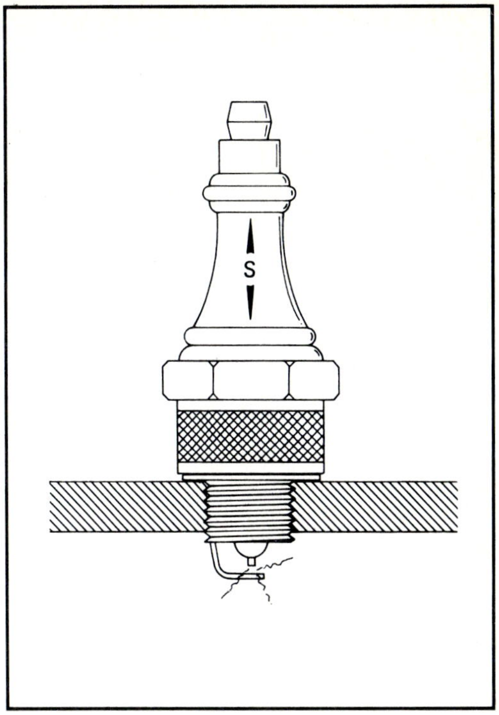

GENERAL

As we have noted before, in order to release the stored energy in fuel, two factors must be present: heat and air. Chapter 5, "Fuel Systems" examined the different ways to mix air and fuel. This chapter looks at methods for providing the heat needed to ignite the air-fuel mixture.

SUMMARY OF COMBUSTION PROCESS

Fuel combustion starts when a high voltage surge of electricity jumps across the electrode gap at the tip of the spark plug. The hot arc ignites the nearby particles of fuel, which, in turn, cause a wavefront of flame to spread throughout the combustion chamber. In a fraction of a second, the air and fuel are converted into a hot mass of expanding gas. The gas pushes on the piston, which transfers the fuel's energy into mechanical force acting on the piston rod and crankshaft. (See Figures 7-1 and 7-2.)

ROLE OF IGNITION SYSTEM

The ignition system's role in the combustion process is to create a high-voltage surge and then deliver it to the spark plug at just the right moment. Therefore, the ignition system must provide (or be provided with) a source of electrical current, and it must time the moment the current appears at the tip of the spark plug. The first part of this chapter will look at the source of electricity for the ignition system.

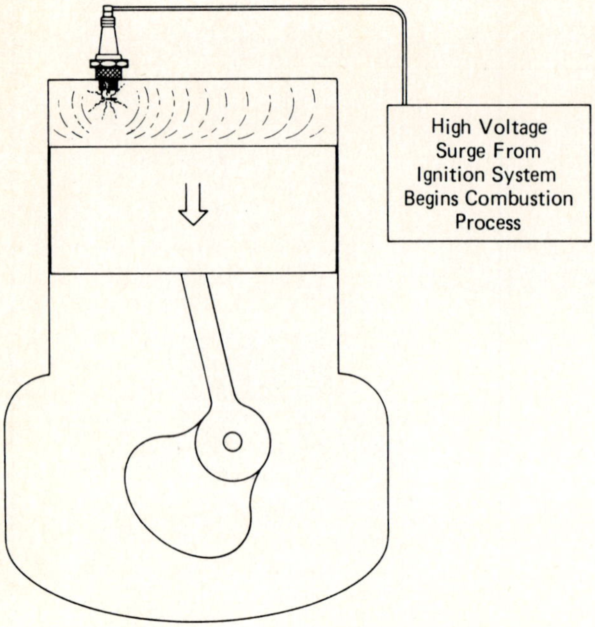

FIGURE 7-1. **Combustion schematic.**

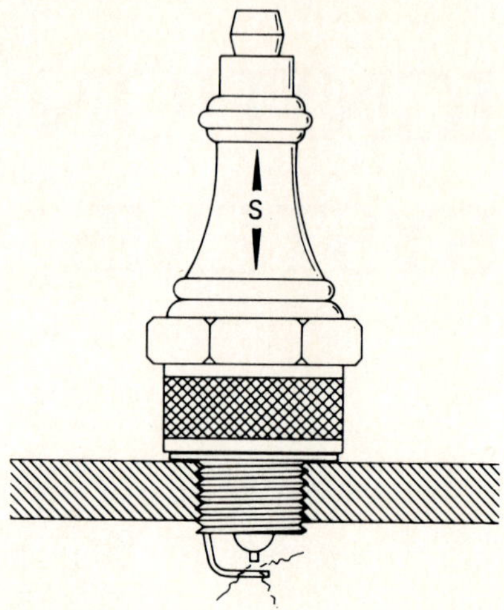

FIGURE 7-2. **Typical spark plug.**

BATTERY-SUPPLIED IGNITION SYSTEMS

All small engines use the magnetic induction principles introduced in the last chapter to produce electrical current. In larger engines with heavier current demands these principles are often employed by a generator (or alternator). The generator supplies charging current to a battery, which in turn provides current for the electrically operated devices associated with the engine (lights, self-starter, horn, instruments, etc.). The battery also supplies current for the ignition system. (See Figure 7-3.)

MAGNETO-SUPPLIED IGNITION SYSTEM

In many small engines, the only electrical component is the ignition system. Such engines do not require a battery or generator. They use the much simpler magneto.

Although magnetos come in various shapes and sizes, most have two components in common—a permanent magnet and a coil (or coils) of wire. Either the coil or the magnet is held stationary while the other component is rotated nearby. Voltage is induced when the lines of force from the magnet intersect the coil's windings. The current produced in the coil is used to supply the ignition system.

TWO KINDS OF MAGNETOS

Magnetos are usually grouped into two main categories: (1) those that are incorporated into some other part of the engine and (2) those that are self-contained.

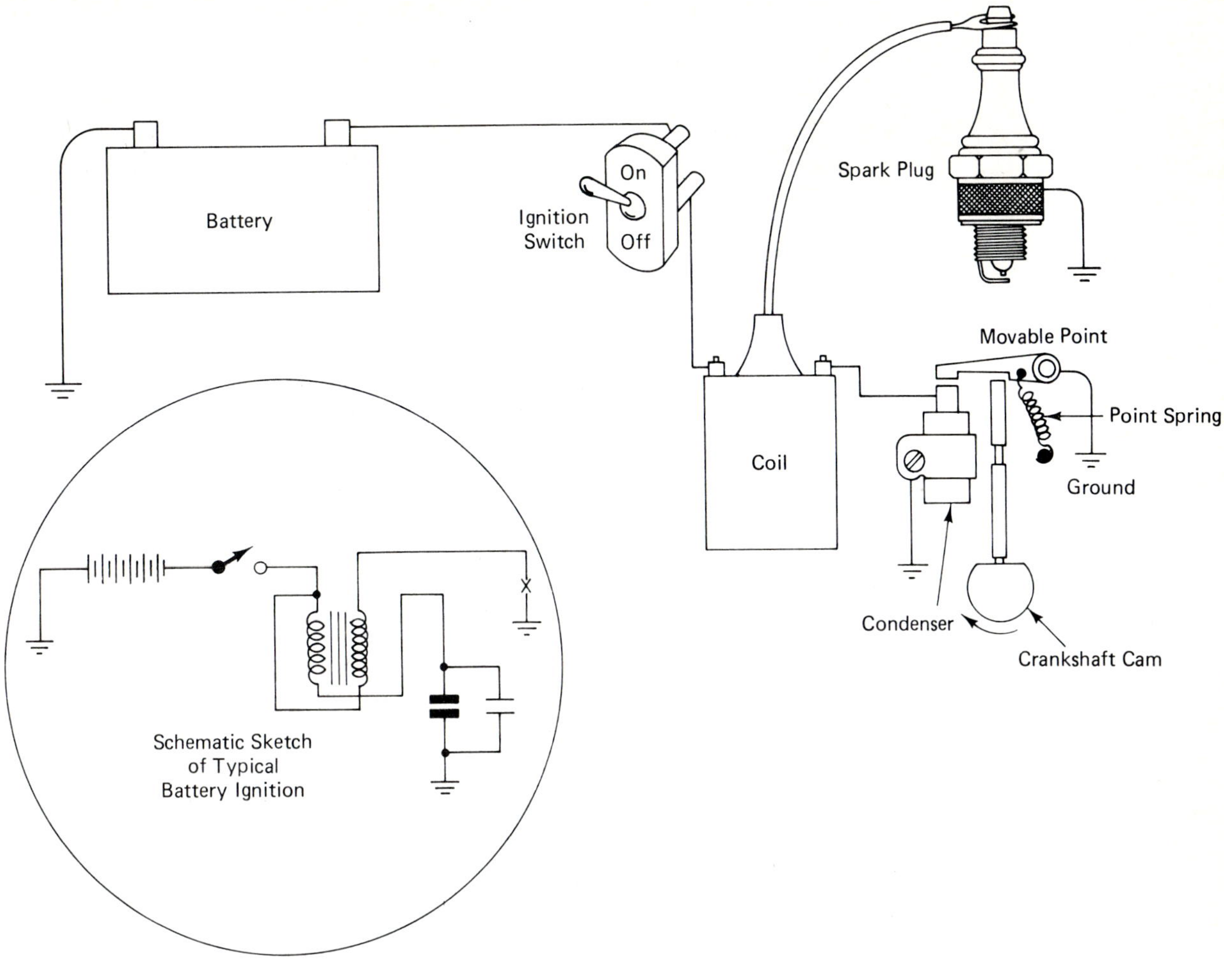

FIGURE 7-3. A typical battery-supplied ignition system.

Self-Contained Magnetos

Self-contained magnetos are usually driven by a gear connected to the camshaft. Some systems rotate the magnet; others rotate the coil. [See Figure 7-4(a), (b), and (c).] In both cases the output fluctuates as the coil and magnet move in relation to one another. For this reason, a self-contained magneto must be "timed" during installation. In other words, the drive gears must be lined up so that the coil and magnet produce peak voltage when the piston is in the optimum position for combustion to begin.

Self-contained magnetos are relatively commonly in "larger," multicylinder small engines. They are often associated with an automotive style distributor which routes the current output to the appropriate cylinder at the appropriate time.

Flywheel Magnetos

By far the most typical technique for producing electricity in single-cylinder small engines is to incorporate the magneto into the flywheel. This is less complicated than self-contained systems. It is also less expensive, since an existing component (the flywheel) is modified to perform multiple tasks.

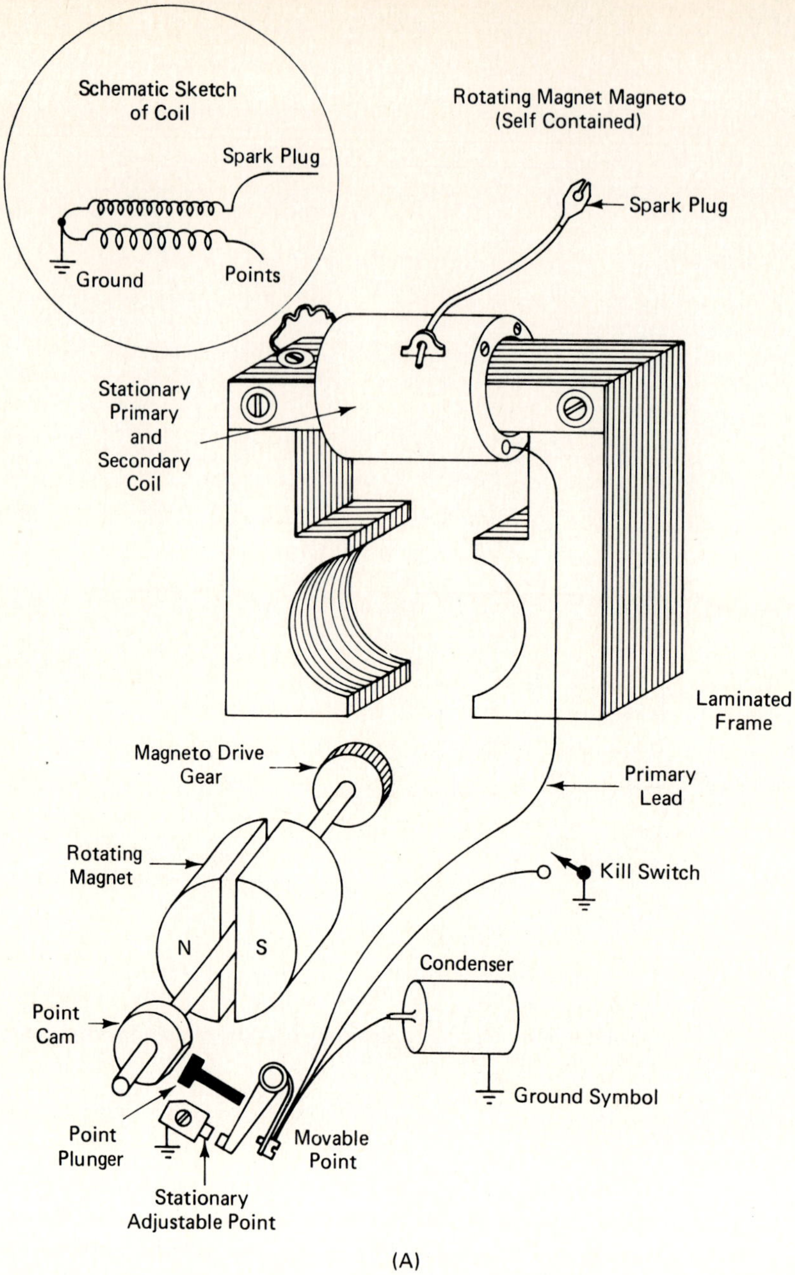

FIGURE 7-4. **(A) Self-contained magnets with rotating magnet; (B) self-contained magnets with rotating coil.**

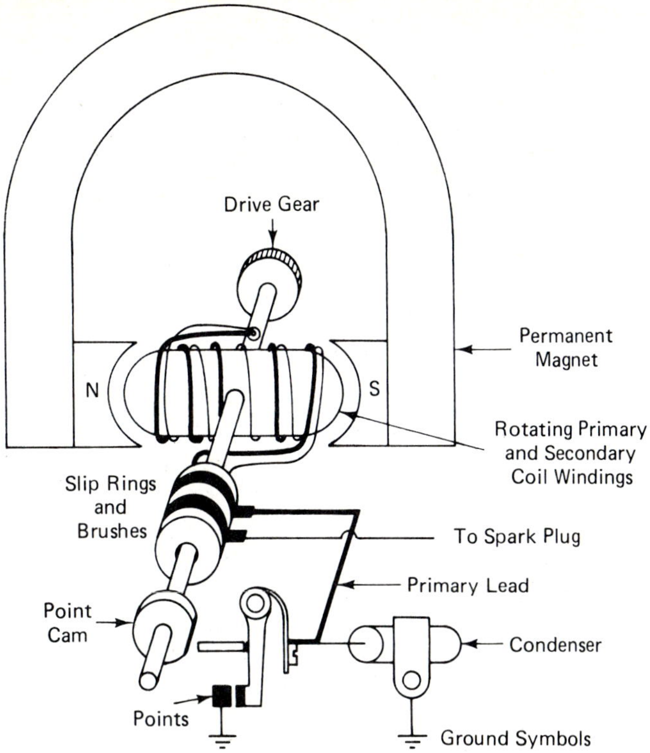

(B)

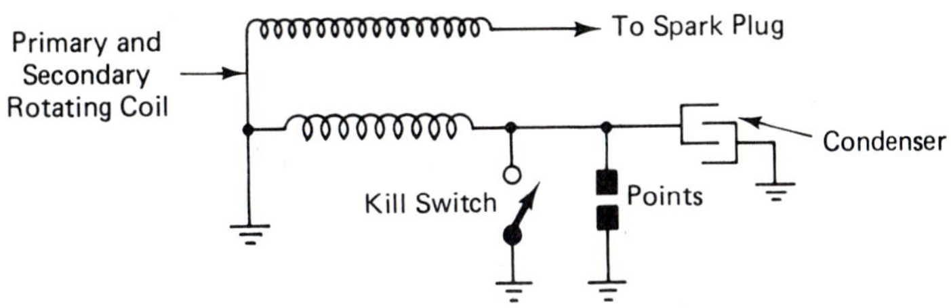

(C)

FIGURE 7-4. (Continued)

In the flywheel system, the permanent magnet is embedded in the outer or inner circumference of the flywheel. As the flywheel rotates, the magnet, which lies flush with the flywheel surface, also rotates.

The coil is positioned near the flywheel – on the inside or outside diameter, depending on the location of the magnet. When the flywheel turns, the lines of force from the rotating magnet cut through the coil and induce voltage in its windings.

Note: Actually, the process is not quite that simple. As you will notice in Figure 7-5, the coil is wrapped around a laminated core piece whose ends lie within a few thousandths of an inch of the flywheel. When the ends of the flywheel magnet line up with the core piece, the lines of force travel through the core and balloon around the entire coil. As the permanent magnet moves away from the core piece, the surrounding lines of force recede from the coil. It is this action that causes current flow through the coil.

We now seem to have identified the source of electricity for the ignition system. Output from the magneto coil is routed directly to the spark plug every time the voltage builds up to a peak. Right? No, not quite.

The problem is that the voltage output from the magneto is not great enough to push electrons across the spark plug's electrode gap. Approximately 20,000 volts are needed; a single magneto coil does not produce nearly this much voltage.

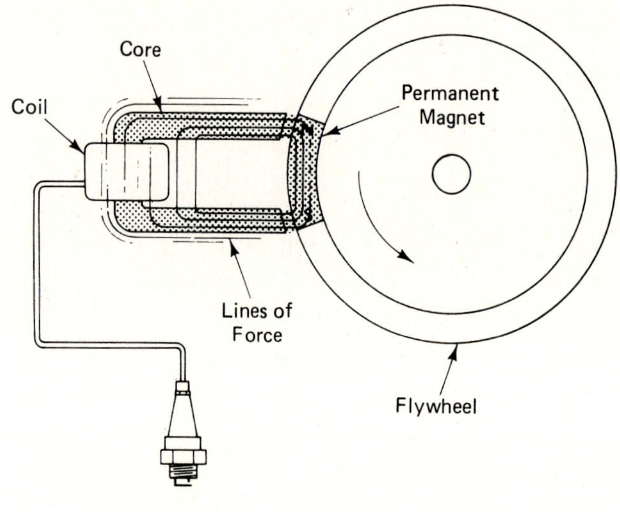

FIGURE 7-5. Schematic of a flywheel magneto system.

IGNITION COIL TRANSFORMER

How, then, is the additional voltage produced? The magneto coil is incorporated into an ignition coil transformer. (See Figure 7-6.) Here is how the transformer works: As noted in the previous chapter, magnetic lines of force circle around any wire through which current flows. Therefore, every time the permanent flywheel magnet induces current flow in the magneto coil, magnet lines of force circle the entire coil. Now, suppose we place another coil inside the first. We will call it the secondary coil and the first one the primary coil. Let us also suppose that a switch is wired into the circuit attached to the primary coil. We will call this switch the points assembly.

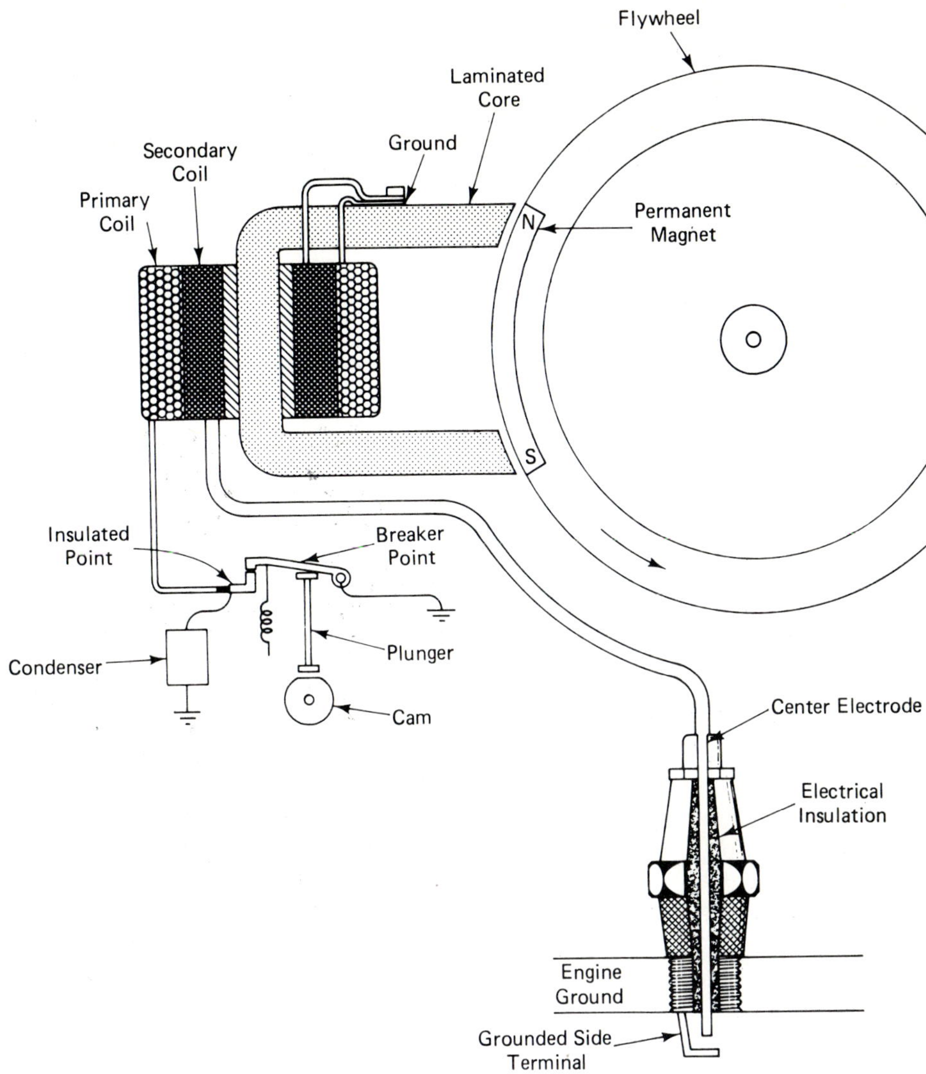

FIGURE 7-6. Schematic of an ignition coil transformer.

Whenever the points open, current flowing through the primary coil is interrupted.

What do you imagine happens to the magnetic lines of force circling the primary coil when the points open? Without current flow, the lines of force cannot exist, so they fall inward to the center of the coil assembly. As the lines of force collapse, they cut across the windings of the secondary coil. Speculating further, what do you think happens then? Naturally, voltage is induced in the secondary windings.

As you have probably already guessed, the output from the secondary is used to fire the spark plug. Sufficiently high voltage is produced because of the difference in the number of windings in the secondary coil compared to the primary coil. The secondary has more turns of wire than the primary, causing it to produce higher voltage. (The total energy output of the secondary is the same as the primary, since the secondary amperage is reduced in proportion to the voltage increase. Energy is neither created nor destroyed by the transformer coil; it is simply rearranged into a more suitable form for igniting the fuel mixture.)

Both the primary and secondary windings are wrapped around the same core piece. Sometimes the primary windings are on the outside; sometimes they are on the inside. The wires that make up the primary and secondary windings are electrically insulated by a sprayed-on coating of varnish or shellac. Because the coating is clear, the wires may appear at first glance to be bare or uninsulated. The entire coil is wrapped in insulating tape or dipped in epoxy.

The relatively limited number of turns in the primary windings are made up of medium size copper wire. One end of the primary wire is grounded to the engine, usually under a small self-tapping screw attached to the laminated iron core piece. The other end is connected to the insulated side of the ignition points.

The relatively greater number of turns in the secondary winding are made up of smaller size copper wire. One end of this wire is also grounded, usually under the same screw used to ground the primary wire. The other end of the secondary wire is attached to a large, heavily insulated cable that goes to the spark plug. [See Figure 7-7(a), (b), and (c).]

Now we have seen how a magneto-supplied ignition system supplies the current used to operate the spark plug. The procedure is similar for a

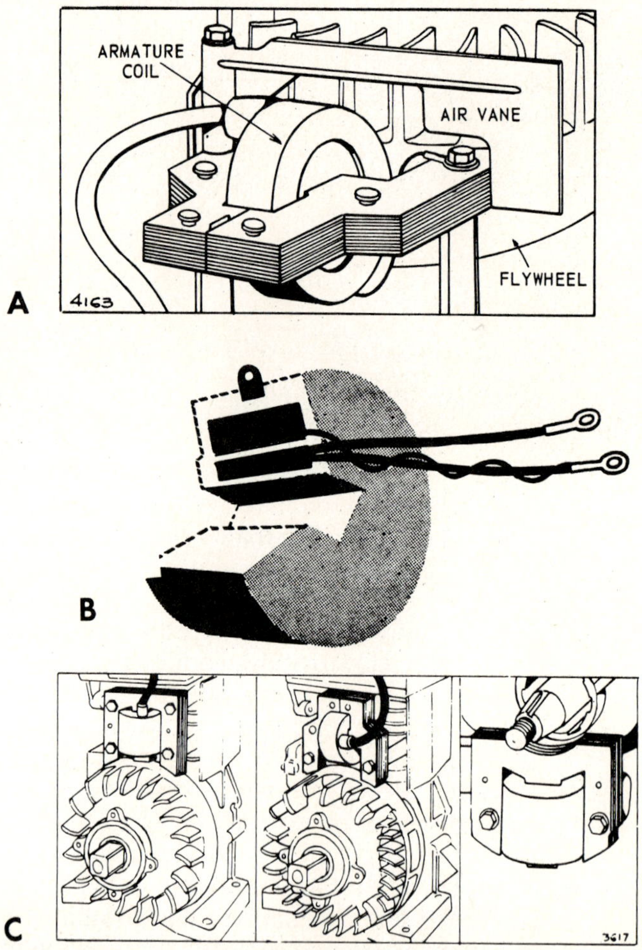

FIGURE 7-7. (A) Pictorial view of an ignition coil transformer; (B) cutaway of an ignition coil transformer; (C) various ways to position a coil transformer (Courtesy of Briggs & Stratton Corporation and Clinton Engines Corporation).

battery-supplied system. Low-voltage current from the battery is delivered to the primary windings of a stand-alone coil. When the primary circuit is interrupted, high voltage from the secondary coil windings is delivered to the spark plug (or plugs).

TIMING AND CONTROL

Next we need to look more closely at the devices that trigger the high-voltage surge from the coil's secondary circuit. Two principal methods are employed—the points systems already mentioned and the solid state, electronic control system.

POINTS TYPE OF IGNITION CONTROL

The points assembly is a mechanical switch wired into the magneto coil's primary circuit. Whenever the points open, the current flow through the primary is interrupted, thereby triggering the high-voltage surge from the secondary coil. When the points are in contact, the circuit is complete, and current can flow through the primary.

Usually, one of the points is stationary and the other is mounted on a pivot arm. The stationary point is referred to as the *insulated point* and the moving point is referred to as the *breaker point*. Point opening occurs whenever the breaker point moves away from the insulated side of the assembly. [See Figure 7-8(a) and (b).]

The points must open every time the piston moves near TDC in the compression stroke. Since piston movement is directly related to crankshaft or camshaft rotation, one of these components is normally employed to actuate the points. A high or low surface is formed on the rotating shaft (cam or crank). This cam surface operates directly on the breaker point or through a rod connected to the breaker point. Every time the shaft completes one rotation, the cam surface causes the breaker point to move away from the insulated point. After the cam surface passes by, the points close again, so the primary circuit is complete once more. In most cases, the points are closed by a return spring. (See Figures 7-9, 7-10, and 7-11.)

Also included in the point assembly is a condenser. It is an electrical device connected to the insulated side of the points. The condenser prevents primary current from arcing across the points as they begin to open (when the point gap is small and easily jumped by low-voltage current). If arcing did occur, the point surface could become pitted and subject to premature failure. The condenser protects the points by acting as a sort of electrical reservoir, storing the primary current until the points are fully open, then discharging the current to ground. [See Figure 7-12(a) and (b).]

The point assembly can be located in various places. Some flywheel magneto ignition systems position the point assembly and coil underneath the flywheel. Other systems place the points and coil outside the flywheel. (See Figure 7-13.)

Two additional topics to discuss before we turn to electronic control systems are: (1) shutoff procedures and (2) timing advance.

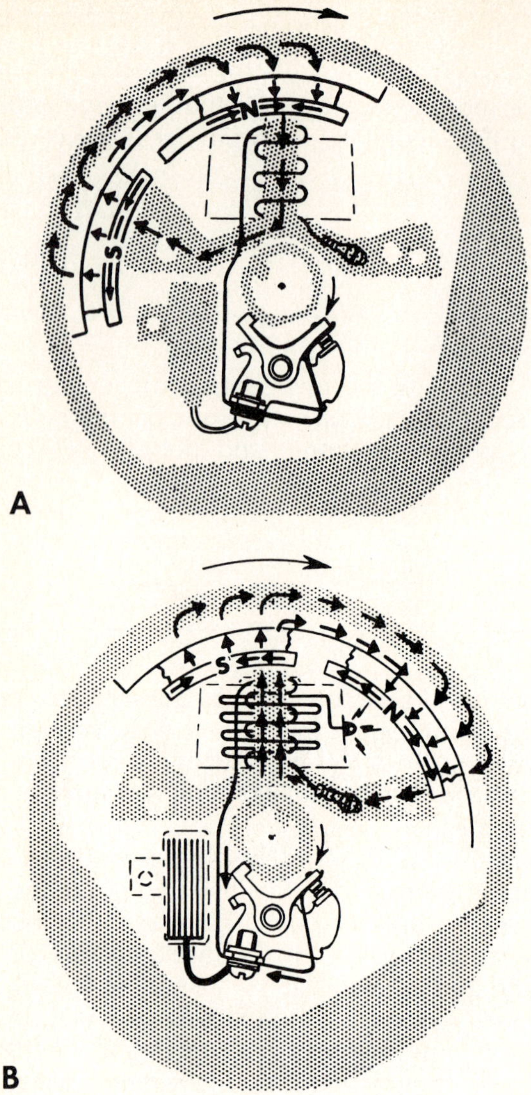

FIGURE 7-8. (A) Current flow with points closed; (B) current flow with points open (Courtesy of Clinton Engines Corporation).

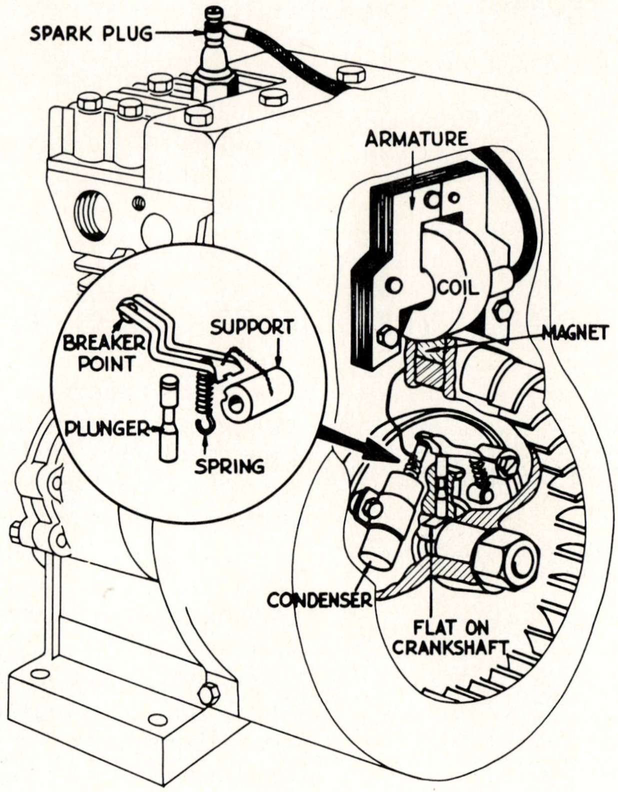

FIGURE 7-9. Point assembly (Courtesy of Briggs & Stratton Corporation).

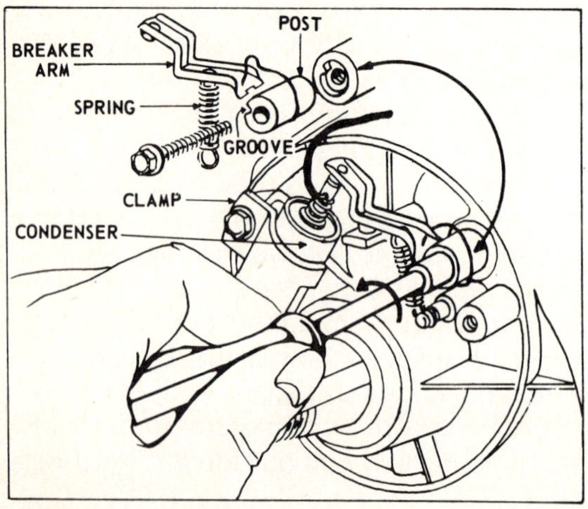

FIGURE 7-10. Adjusting the points (Courtesy of Briggs & Stratton Corporation).

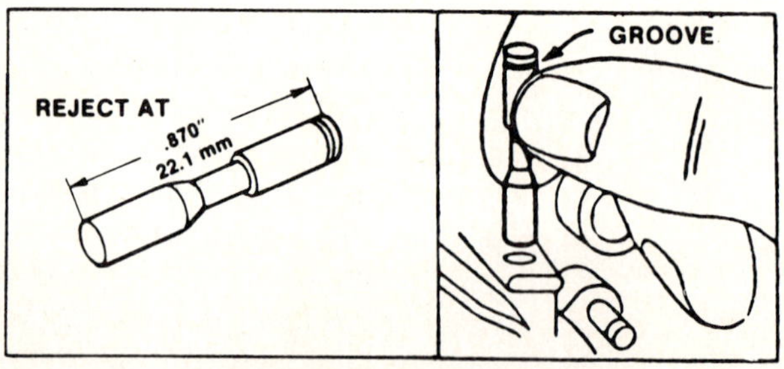

FIGURE 7-11. Breaker point operating plunger (Courtesy of Briggs & Stratton Corporation).

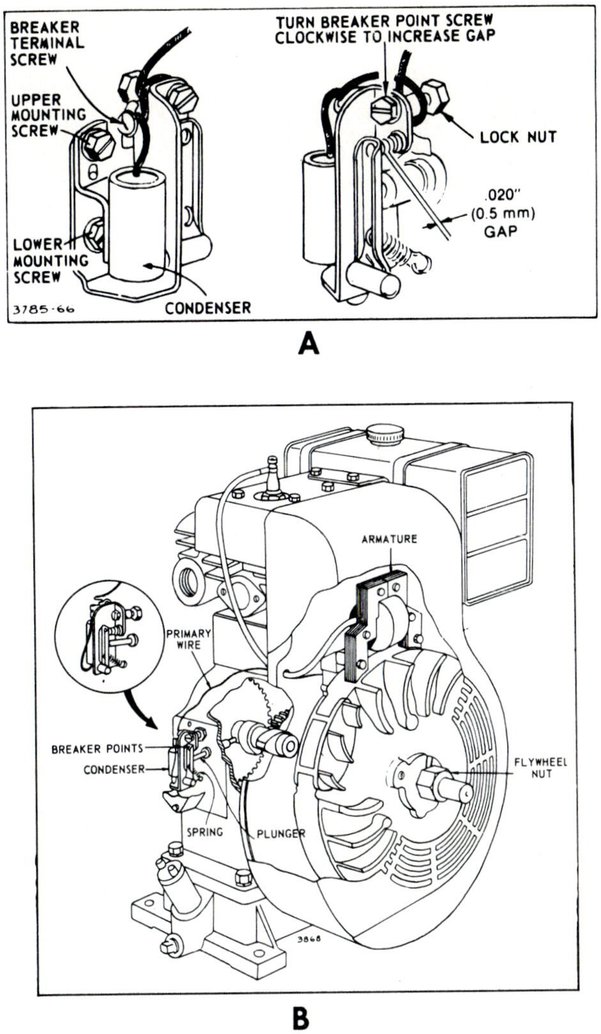

FIGURE 7-12. (A) Condenser; (B) condenser location (Courtesy of Briggs & Stratton Corporation).

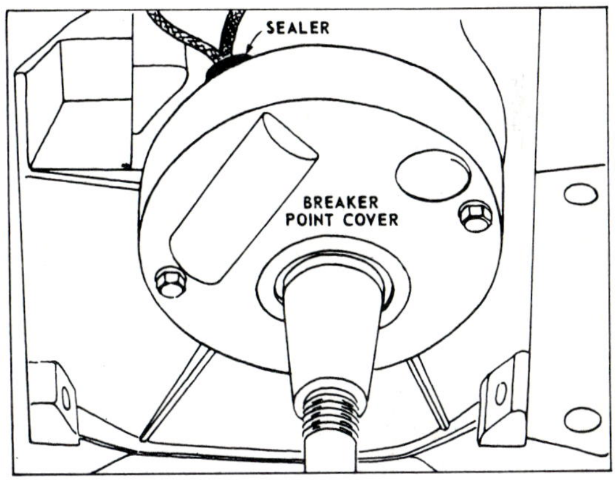

FIGURE 7-13. Breaker point cover (Courtesy of Briggs & Stratton Corporation).

SHUTOFF PROCEDURES

The most effective way to shut off an engine is to interrupt the operation of the ignition system. In battery-operated systems, an ignition switch is wired into the low-voltage primary circuit. Opening the switch prevents current flow.

Magneto-supplied ignition systems are often shut off by grounding the points. In other words, the insulated side of the points is connected to ground so that the circuit remains complete all the time, even when the points are open. In that way, no voltage can be induced in the secondary.

To see how this works, consider the shutoff system used by a common push-type lawn mower. A wire is passed from the insulated point to a grounding switch. This switch, consisting of a knife bladelike ground connector, is connected to the throttle cable. When the throttle is pushed to the OFF position, the knife is moved into ground contact. Then, when the throttle is pulled ON, the blade is moved away from ground so that the point circuit can function in a normal manner.

TIMING ADVANCE

The fuel burning process, once initiated, tends to proceed at a uniform rate, regardless of engine speed. Consequently, there is less time for combustion when the engine is running faster. Engines used in automobiles and other types of transportation devices usually provide some means for advancing the spark as the engine speeds up. In other words, instead of interrupting the primary circuit when the piston is at TDC (or later), the circuit is broken earlier (10 to 20 deg BTDC in some cases).

Up until the era of electronic, computerized control systems, one common way to advance the spark was to attach a set of centrifugal weights to the breaker point control shaft. As the engine speeded up, the weights would swing out, causing the shaft to shift position and open the points earlier. This method is still used on some self-contained magnetos employed in "larger" small engines.

ELECTRONIC IGNITION CONTROL SYSTEMS

Points-type systems work well in lower-speed small engines. Points also work well in higher-speed engines. However, at increased speeds and higher voltage levels, the points tend to wear out more quickly. Worn points can cause misfiring, which in turn can result in reduced power and increased fuel consumption. Consequently, when solid electronic systems became available to control the primary circuit, many manufacturers of small engines were quick to adopt the new technology. Not only do the solid state systems last longer, they also tend to produce a higher-voltage spark—which is especially important when one is hand cranking a potentially balky engine.

A typical flywheel-based electronic system includes several of the components used in points-type systems, the major ones being (1) a permanent magnet embedded in the flywheel and (2) an ignition transformer coil. As before, the field from the permanent magnet is the primary energy source for the coil. Also, as before, a high-voltage surge comes from the secondary windings whenever lines of force collapse around the coil assembly. However, the components for accomplishing these actions are altogether different. Several new elements are used, including a solid state diode and transistor.

Without getting too deeply into the operation of these solid state semiconductor devices, we can simply say that they function as valves for controlling current flow. Diodes act like one-way valves, allowing current to flow in only one direction through a circuit. Transistors are used as switching valves. A weak current flow at the transistor base can control a much higher-voltage current through the transistor emitter and collector. If no current flows through the base, flow from the emitter to the collector is blocked altogether. (See Figure 7-14.)

One other electrical component that should also be mentioned is the capacitor. (See Figure 7-15.) It is not a solid state device and was in general use long before the introduction of transistors and diodes. Capacitors act as temporary storage locations for electrical charges. Current flowing into a capacitor builds up inside before discharging as a sudden burst of accumulated electrons.

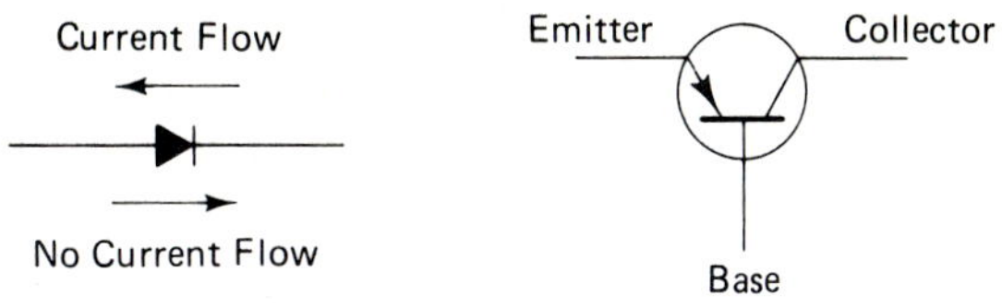

FIGURE 7-14. **Diodes and transistor.**

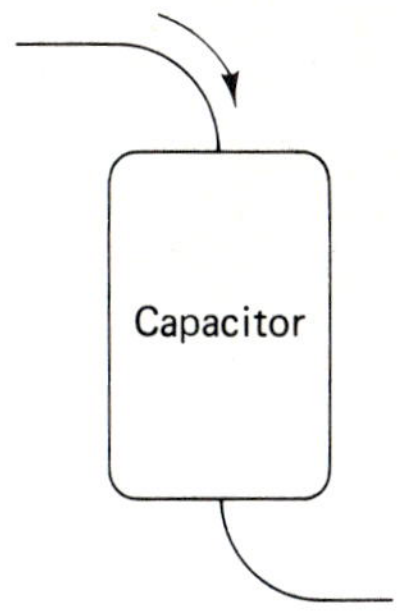

FIGURE 7-15. **A capacitor.**

Now refer to Figure 7-16 and the following paragraphs to see how a simplified solid state ignition control system works:

1. As the flywheel rotates, the embedded permanent magnet passes by a charging coil.
2. The lines of force from the permanent magnet induce voltage in the charging coil. The voltage pushes electrons through a diode to the capacitor. A charge of electrons builds up in the capacitor.
3. As the flywheel continues to rotate, the permanent magnet passes by a second coil, called the *trigger coil*. Current created in the trigger coil is directed to the base of a transistor.
4. Current flow in the base circuit switches the transistor ON. The charge from the now saturated capacitor is able to escape through the transistor

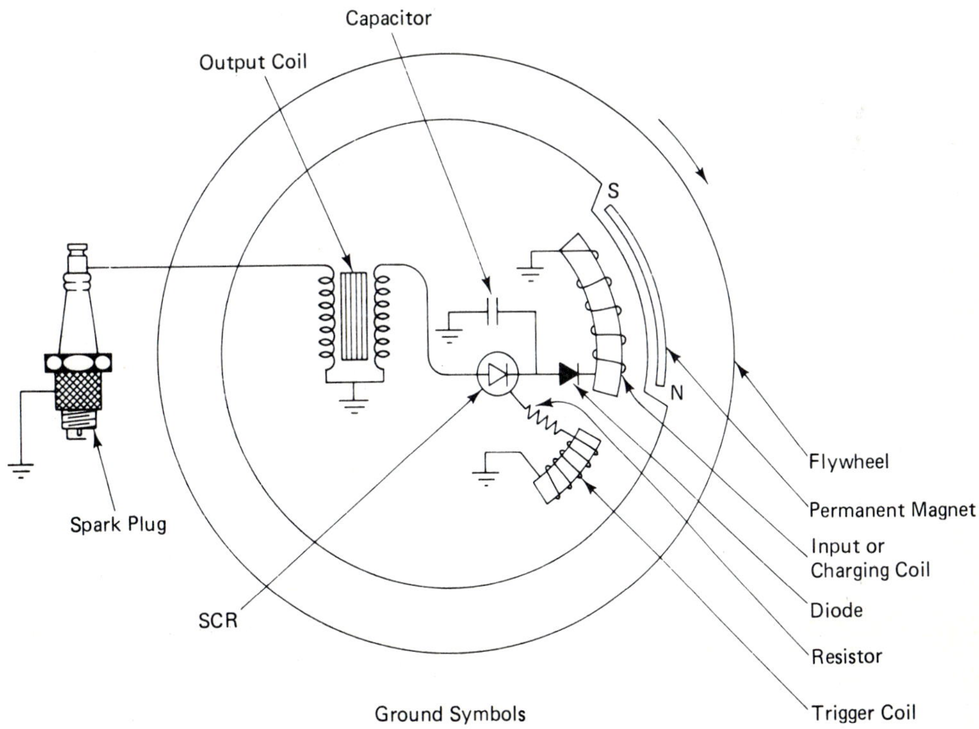

FIGURE 7-16. **A typical solid-state ignition control system.**

emitter/collector to the primary coil windings. (The charge cannot go back the other way because of the one-way diode.)

5. The momentary surge of current flow from the capacitor builds up a magnetic field around the coil. When the surge dissipates, the field collapses across the secondary windings.
6. The lines of force cutting across the secondary windings induce the high-voltage surge needed to fire the spark plug.

8

CRANKING AND CHARGING SYSTEMS

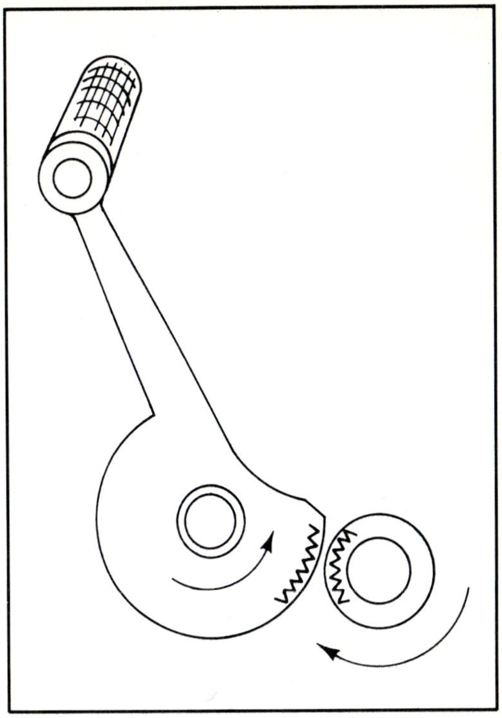

GENERAL

This chapter examines (1) manual cranking systems used to start small engines, (2) electrical self-starters, and (3) the generating systems associated with electrical self-starters.

REQUIREMENTS

The main object of any cranking system is to rotate the engine long enough and fast enough for it to begin running under its own power. A secondary objective is to disengage the cranking mechanism after the engine is running. To be effective, the starting system must be able to overcome two forces: (1) the inertia (resistance to movement) of the engine's rotating components and (2) the compression developed by the piston as it moves up in the cylinder.

In the beginning, all internal combustion engines used manual cranking systems. In other words, human muscle power had to overcome engine inertia and compression. Then, as battery and electrical technology developed, most larger engines were provided with electrical self-starters. Improvements in starting systems were a principal factor in promoting the widespread use of internal combustion engines. After all, if only a large strong person could crank an engine, its use was bound to be limited.

Today, many "larger" small engines, such as those powering self-propelled garden tractors, employ electric self-starters. Some "smaller" engines also use electric self-starters. However, the majority of small engines are still likely to be equipped with manual starters. Consequently, we will begin our discussion by examining the different types of manually operated systems.

HAND CRANKS

Many early small engines took a very direct approach to cranking. Operators simply grabbed a large, exposed flywheel, turning it until the engine started running. This method was usually difficult and occasionally hazardous. So, before long, hand cranks became common.

A hand crank resembles a lug wrench. (See Figure 8-1.) The operator places one end over the exposed tip of the crankshaft, then turns the other end. The end placed over the crankshaft is somewhat socket shaped, with a sloping notch cut into one side. The notch fits over a pin sticking out at a right angle from the end of the crankshaft. Because of the notch, the hand crank can turn the crankshaft, but not vice versa. So, after the engine starts running, the hand crank is thrown out of engagement with the crankshaft. Depending on the size of the engine and other factors associated with the cranking operation, the disengagement process can be quite sudden and violent. Unaware or unlucky operators were sometimes injured as a result.

ROPE AND SPOOL STARTERS

Hand cranks were commonly used to start larger engines, like those found in early automobiles. Hand cranks were also used on many small engines, since most of these engines were not really small. However, as technology advanced to the point where usable amounts of power could be extracted from truly small power plants, the clumsy hand crank was not necessary. Other, more convenient cranking methods could be employed.

One of the first alternatives to the hand crank was the rope and spool starter. (See Figure 8-2.) The spool resembles a large pulley with a wide flat groove. It is connected to the crankshaft, opposite the power takeoff end. Like a hand crank, the upper lip of the pulley contains a notch with one sloping side.

To start the engine, the operator first places the knotted end of the pull rope on the outside of the notch. Then the rest of the rope is wound around the flat side of the spool. Once the rope is in place, a "T" handle on the other end is pulled vigorously. This action spins the engine until the rope unwinds and the knot comes out of the notch. Assuming the operator

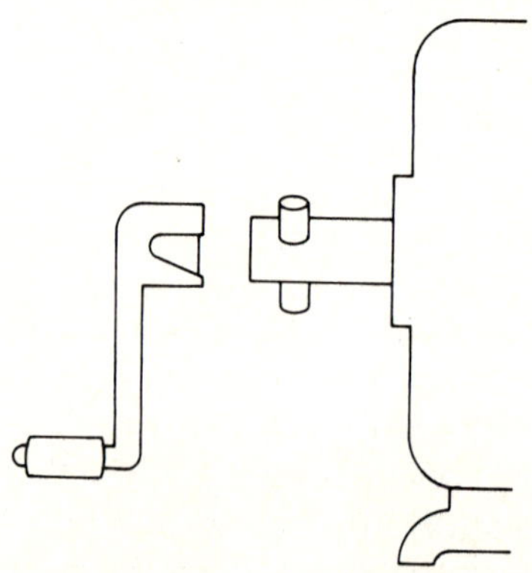

FIGURE 8-1. A hand crank.

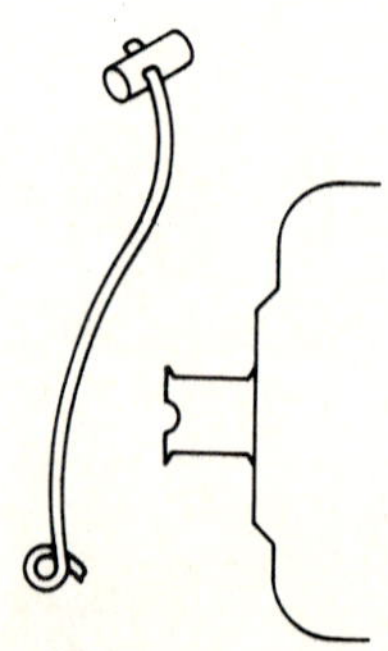

FIGURE 8-2. A spool and rope crank.

is lucky and everything is working okay, the engine will probably crank after two or three tries.

Rope and spool starters were used for many years to crank small engines. However, there were some problems. The rope was easily lost and the exposed, rotating spool presented a hazard to the unwary user.

FOOT/KICK STARTERS

As we noted earlier in this book, some early washing machines were driven by small, air-cooled engines. These units were especially popular with housewives in rural areas that had not been supplied with electric power. To make these washing machines more attractive, some manufacturers provided foot-operated kick starters, as opposed to the then conventional rope and spool starters. The kick starter included a small gear and ratchet assembly mounted on the end of the crankshaft and a larger, quadrant gear attached to the kick starter. When the pedal was pushed down, the quadrant gear turned the crankshaft gear, which spun the engine over. After the foot pedal was released, a return spring brought the quadrant and pedal back into a "park" position, out of mesh with the crankshaft gear. (See Figure 8-3.)

This method of cranking never did become very popular with manufacturers of general-purpose small engines. However, kick starters are still widely used to crank motorcycle engines.

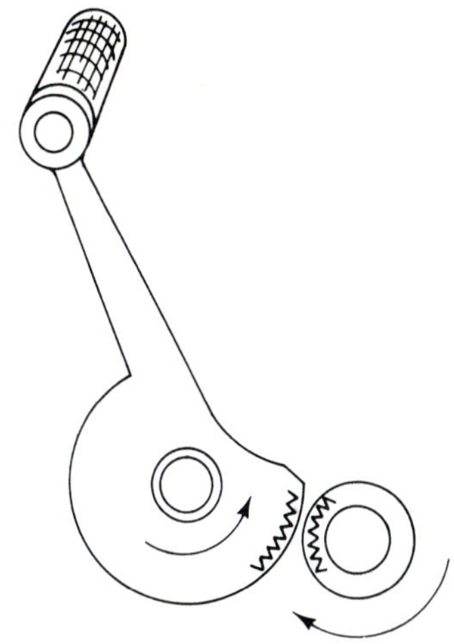

FIGURE 8-3. A kick starter.

RECOIL STARTERS

The recoil starter is an adaptation of the rope and pulley starter with some important improvements. The pulley spool is attached to the crankshaft by a ball-type, one-way clutch. Also included is a flat return spring assembly. When the rope is released, after being pulled out by the operator, the spring automatically rewinds the rope back around the spool. The walls of the spool are also different, formed into a V shape so that the turns of rope lie on top of one another (instead of side by side, as in a flat-side rope and spool starter). The handle end of the rope passes out through the starter cover, which is formed integrally with the flywheel shroud. As you are probably aware, many small engines still use the recoil starter. (See Figure 8-4.)

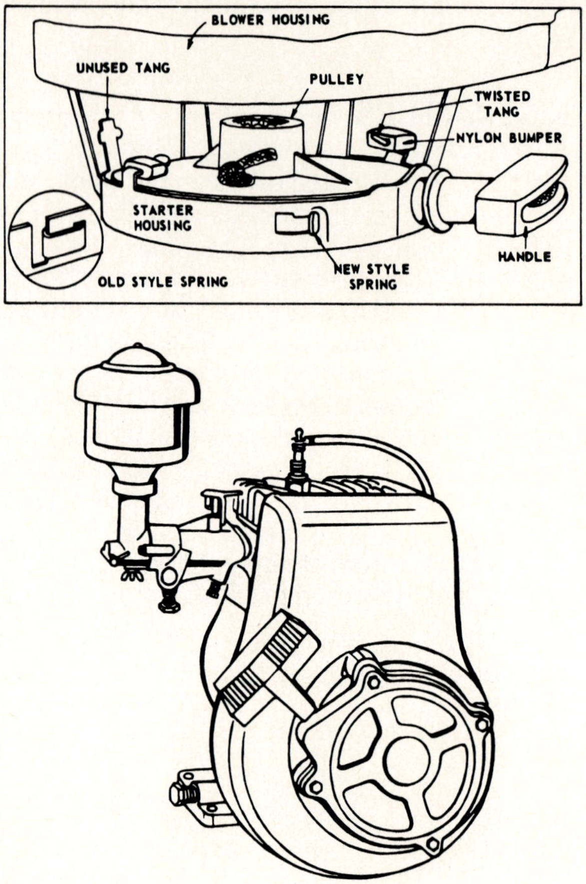

FIGURE 8-4. Recoil on a rewind starter (Courtesy of Briggs & Stratton Corporation and Clinton Engines Corporation).

WINDUP STARTERS

Windup starters are most common on small engines of the vertical drive shaft design. The mechanism, which is located under the airshroud, includes: (1) a one-way ball clutch, (2) a spring-loaded ratchet and gear assembly, (3) a folding cranking handle, and (4) a release knob. Turning the cranking handle places the spring under load. Then, when the control knob is released, the spring unloads, pushing on the gear and ratchet assembly, which work through the one-way ball clutch to spin the engine. Windup starters do not require the sudden burst of manual exertion needed to pull recoil starters and in this respect are easier to operate. However, windup starters are still not as widely used as recoil starters. [See Figure 8-5(a), (b), and (c).]

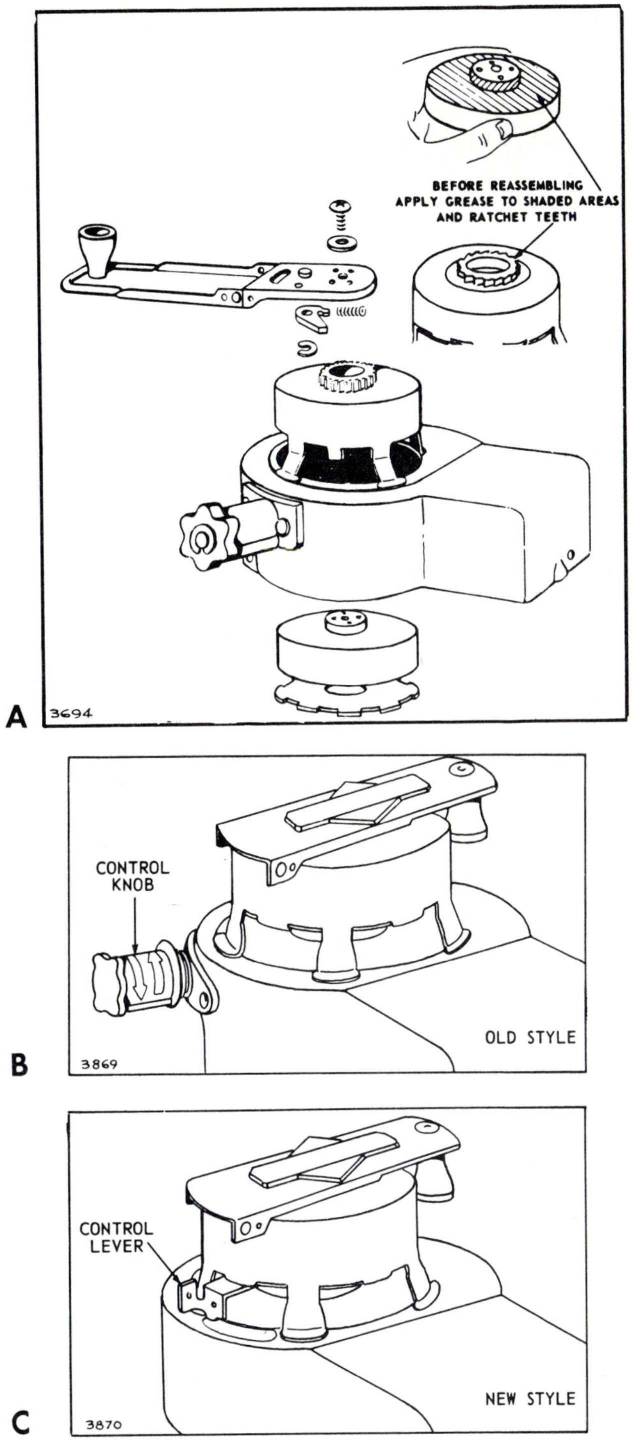

FIGURE 8-5. (A) A wind-up starter; (B) an old-style wind-up starter; (C) a new-style wind-up starter (Courtesy of Briggs & Stratton Corporation).

VERTICAL PULL STARTERS

Vertical pull starters are a variation of recoil-type starters. The main difference is the addition of a small pulley which allows the rope to be pulled up vertically, instead of in a horizontal direction. This is considered to be an easier cranking position by some users.

ELECTRICAL SYSTEMS

The remainder of this chapter examines various kinds of electrically operated starters and associated charging systems. Some of these systems use electrical power just for starting; others supply power for electrically operated accessories. (See Figure 8-6.)

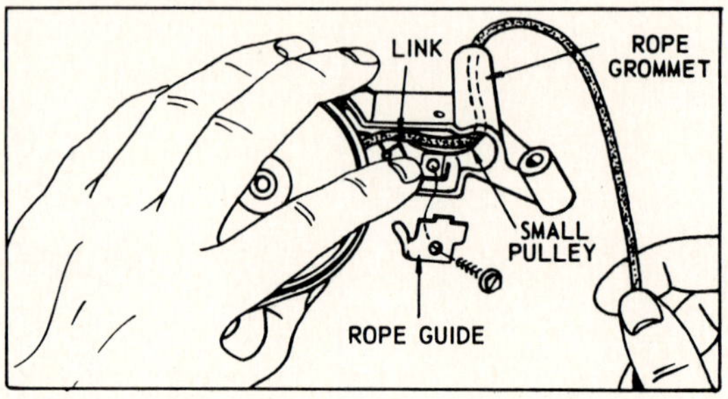

FIGURE 8-6. Installing a pulley on a vertical starter (Courtesy of Briggs & Stratton Corporation).

STARTER GENERATORS

As noted in Chapter 6, it is theoretically possible to operate an electric motor as a generator and vice versa. That is exactly the kind of system employed by some "larger" small engines. The starter/generator is connected to the crankshaft by a V-belt running from a small pulley on the starter generator to a larger pulley on the crankshaft.

The starter/generator contains two sets of field windings, one made up of larger wires and the other of smaller wires. When the starter switch is engaged, power is directed to a solenoid switch which closes a higher-amperage circuit leading to the larger field windings. Current flowing through these windings creates regions of magnetic attraction and repulsion which cause the armature to rotate. Since the armature is fabricated around the starter drive shaft, the drive pulley also rotates to spin the engine.

After the engine is running and the starter switch is released, no current flows to the solenoid. Consequently, the higher-amperage circuit is opened and current also stops flowing to the larger field windings. However, by this time the engine is rotating the armature. Lines of force from the smaller field windings induce voltage in the rotating armature. The resulting current, which is rectified by the commutator into DC (direct current), charges the battery and helps supply any other electrical devices included in the circuit, such as lights, a horn, and a battery-operated ignition system. [See Figure 8-7(a), (b), and (c).]

Batteries used in starter generator systems usually have six 2-volt, lead acid-type cells for a total output of 12 volts. Charging current to the battery is controlled by a voltage regulator.

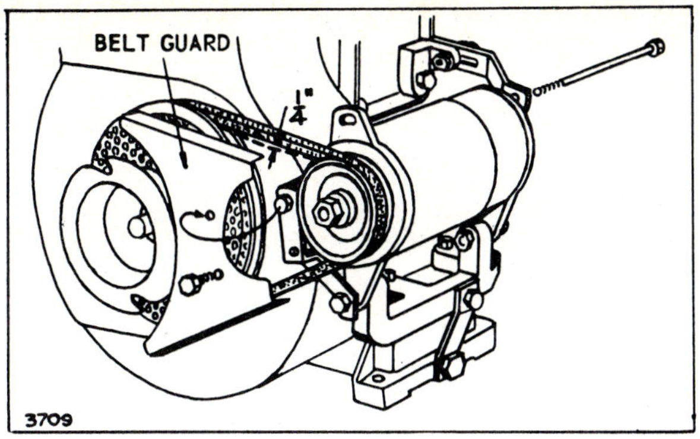

A

B

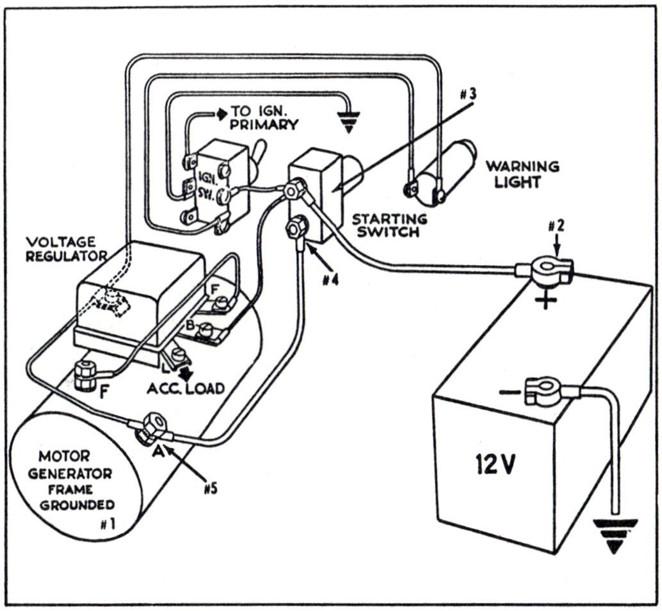

C

FIGURE 8-7. (A) and (B) Twelve-volt starter generators; (C) wiring diagram of a twelve-volt starter generator (Courtesy of Briggs & Stratton Corporation and Clinton Engines Corporation).

GEAR DRIVE STARTERS

A number of small engines use a separate, gear-drive starter motor. These motors, which are about the size of soup cans, are similar to starter motors used on automobiles.

When the cranking switch is turned on, higher-amperage current goes to the field windings. Then as the armature rotates, a pinion gear at the end of the armature shaft turns a large ring gear formed around the flywheel. The pinion gear is moved in and out of mesh with the ring gear by a "bendix" mechanism. Here is how the bendix works:

When the armature begins to rotate, the pinion is screwed down and along helical threads cut into the armature shaft. A locking mechanism holds the gear in place at the end of the shaft while cranking occurs. Then, when the engine starts running faster than the starter motor, centrifugal force unlocks the pinion. It is then driven back down the helical threads to the other end of the shaft and out of engagement with the ring gear. [See Figure 8-8(a) and (b).]

Most gear drive starters used in small engines operate in this manner. However, electrical power for the starter motors is supplied in different ways, depending on the size of the engine and the number of accessories being operated.

Next, we will look at a few of these electrical supply systems.

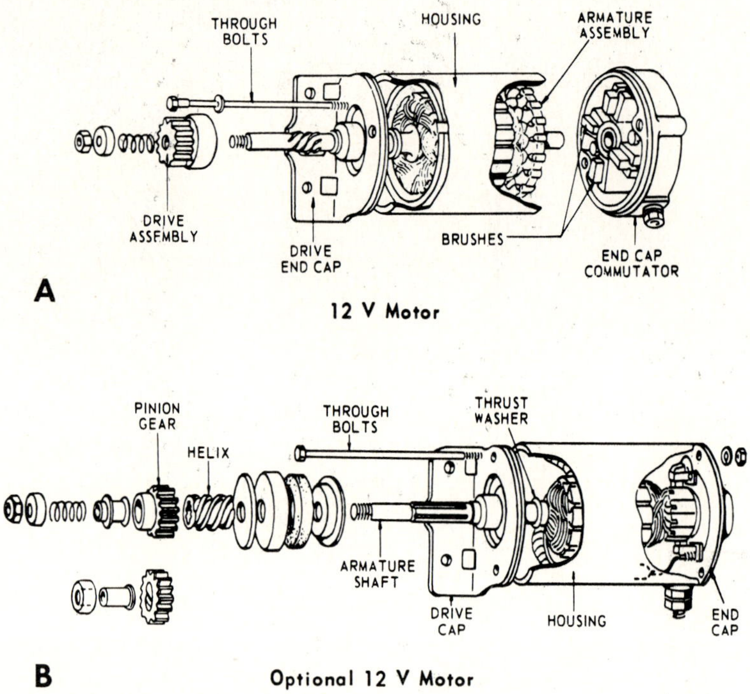

FIGURE 8-8. (A) An optional twelve-volt starter (exploded view); (B) a twelve-volt starter motor (exploded view) (Courtesy of Briggs & Stratton Corporation).

110-VOLT SUPPLY SYSTEM

Some small engine applications that don't require electrical power to operate accessories use ordinary 110V house current to drive a starter motor. The operator simply plugs in the starter motor to the 110V outlet (making sure that a proper ground is present). Then, when the engine is running, the starter motor is unplugged. Because of the higher voltages used, these systems employ heavy-duty insulators to prevent shocks. Also, the motor windings are somewhat different than those used in lower-voltage systems. [See Figure 8-9(a) and (b).]

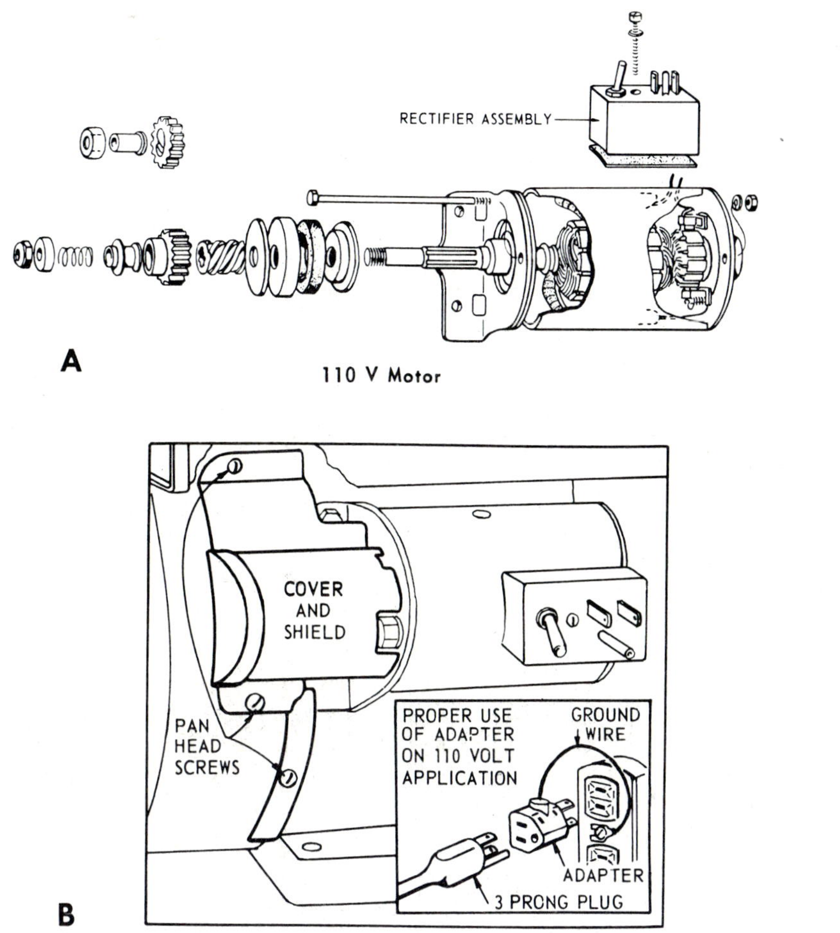

FIGURE 8-9. (A) Exploded view of a 110-volt starter; (B) a 110-volt gear drive starter (Courtesy of Briggs & Stratton Corporation).

TWELVE-VOLT NICKEL CADMIUM SYSTEM

As you might expect, the 110V system is best suited for stationary applications where a supply of 110V current is readily available. Portable small engines, on the other hand, require a self-contained source of electrical power.

"Smaller" small engines (like those found in push-type lawn mowers) sometimes employ a 12-volt nickel cadmium (NICAD) battery. NICAD systems do not usually include any accessories; the battery's sole job is to turn over a small, gear-type starter motor. Nor is an engine-driven charger normally found in these systems. The NICAD battery contains enough power for several starts. When not in use, it is connected to a small battery charger plugged into a 110V outlet. This same type of NICAD charging system is employed by a number of small, portable, electrical devices. (See Figure 8-10.)

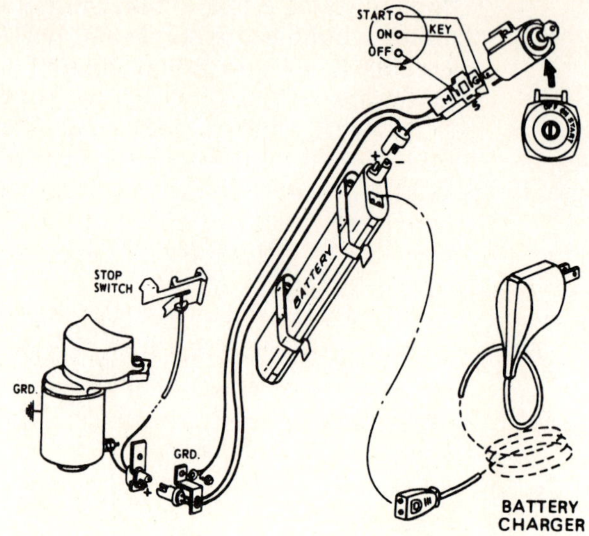

FIGURE 8-10. The wiring diagram of a nickel cadmium starting system (Courtesy of Briggs & Stratton Corporation).

TWELVE-VOLT LEAD ACID TYPE BATTERIES

"Larger" small engines (like those found on riding garden tractors) often use fairly large 12-volt lead acid-type storage batteries which are recharged by an engine-driven generator or alternator. This kind of system is similar to systems used in automobiles. The power produced not only operates a starter motor, but supplies a variety of accessories as well as a battery-operated ignition system.

Since the engines using such systems are not really all that small, we won't go into detail about their fairly complex electrical components. However, we will cover a few basic points to give you an idea of some of the possible variations.

Alternators Versus Generators

Most modern systems employ an alternator to recharge the battery. Alternators perform the same basic tasks as generators; they send current in the opposite direction to normal battery flow to build up a supply of free electrons on the battery's plates. However, alternators are lighter in weight than generators and can be driven at higher speeds to produce a more reliable supply of current. Also, unlike generators, the alternator output comes from the stationary field (stator) windings rather than the rotating armature (rotor) windings. Therefore, a split ring commutator cannot be used to change (rectify) the alternating current into DC flow. Instead, alternators use a system of interconnected diodes. They act as one-way current valves to make sure the current always flows in one direction.

Various types of alternator systems are used on "larger" small engines.

NONREGULATED 1½- AND 4-AMP ALTERNATORS

Like most small engine alternators, these alternators are constructed integrally with the flywheel. A ring of magnets is mounted on the inner surface of the flywheel. Positioned within the flywheel and near the magnetic ring is the stationary stator assembly. When the flywheel rotates, the lines of force from the magnetic ring cut across the stator windings to induce voltage which creates current flow. This AC current is changed by the solid state diode rectifier system into direct current to recharge the battery and help supply the accessories.

One and one-half- and four-amp alternators generally do not use a voltage regulator. Therefore, the output fluctuates as the engine speed changes. Rated output usually occurs at around 3600 RPM. These alternators generally use less than 0.15 horsepower. (See Figures 8-11, 8-12, and 8-13.)

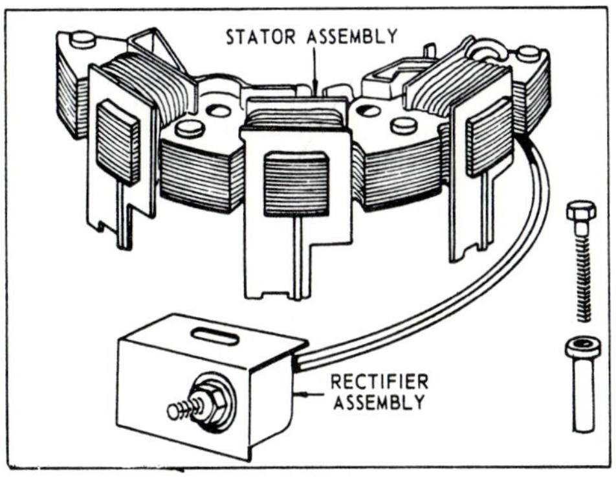

FIGURE 8-11. A one-and-one-half-amp alternator assembly (Courtesy of Briggs & Stratton Corporation).

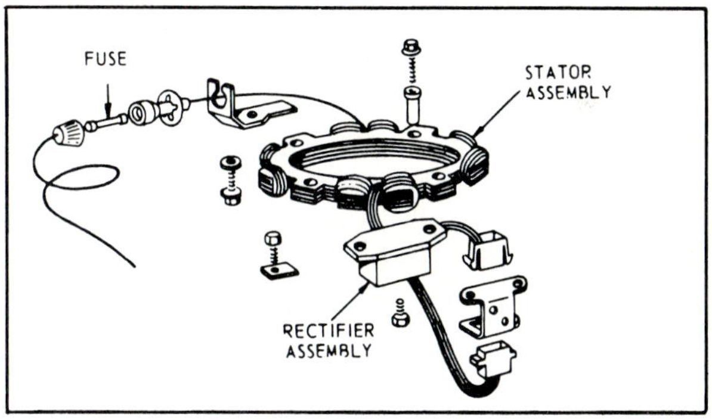

FIGURE 8-12. A four-amp alternator (Courtesy of Briggs & Stratton Corporation).

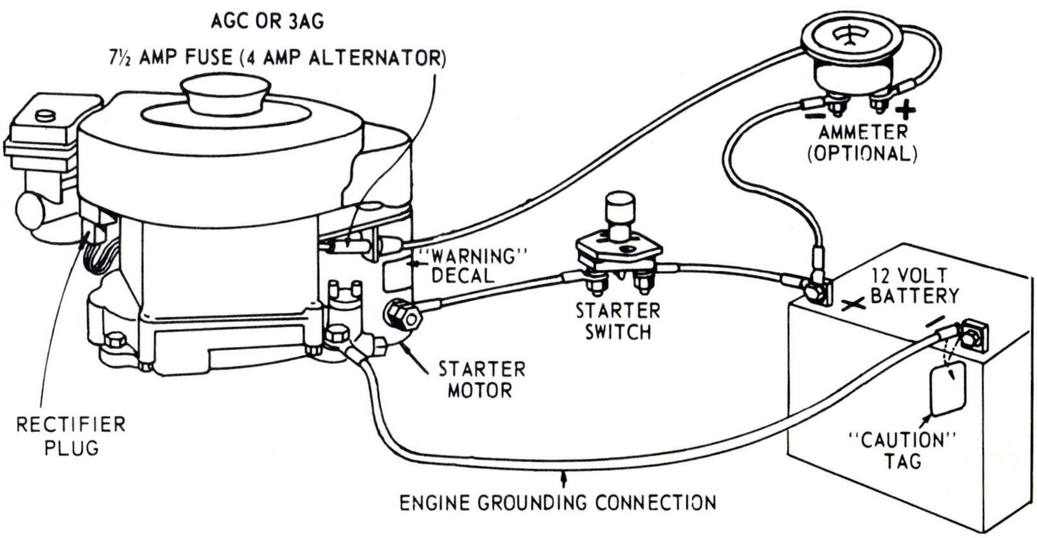

FIGURE 8-13. Typical wiring installation for a four-amp alternator (Courtesy of Briggs & Stratton Corporation).

TEN-AMP REGULATED ALTERNATORS

These alternators, which are used most often on even "larger" small engines, are similar to the lower output 1½- and 4-amp systems. The main difference is the addition of a voltage regulator to control the output within a given range. (See Figure 8-14.)

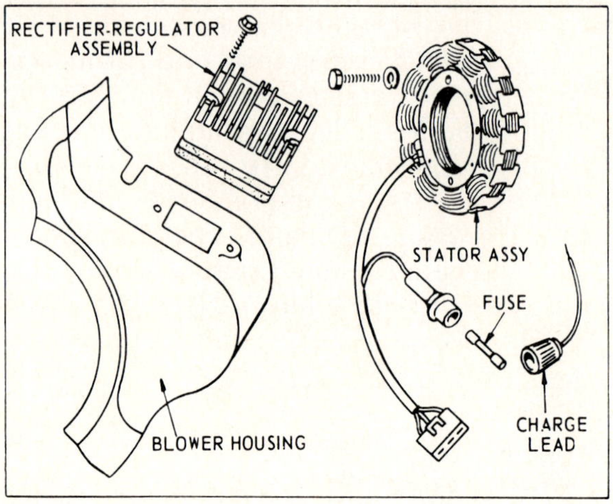

FIGURE 8-14. A ten-amp alternator assembly (Courtesy of Briggs & Stratton Corporation).

DUAL CIRCUIT ALTERNATORS

These chargers, found on some Briggs and Stratton engines, are actually two alternators in one. A single ring of magnets is mounted on the inner surface of the flywheel. This rotor unit supplies magnetic lines of force for two separate sets of stator windings. One set of stator windings is used to recharge a 12-volt storage battery. The battery's sole function is to supply

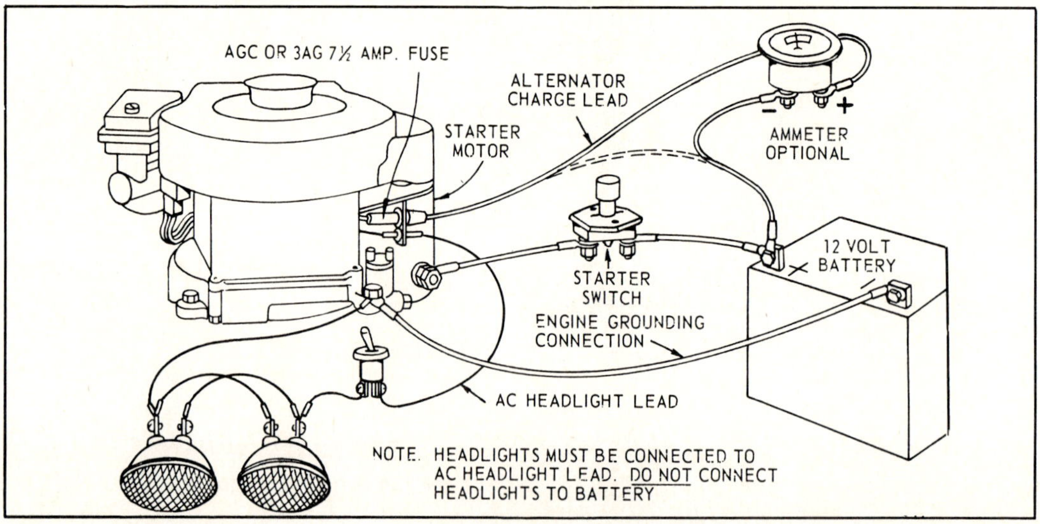

FIGURE 8-15. Typical wiring installtion for a dual-circuit alternator (Courtesy of Briggs & Stratton Corporation).

current for a gear-type starter motor. The other set of stator windings provides power for the lights or any other accessories included in the system. Since the two circuits are not connected, the lights and other accessories do not drain the battery.

The output of the accessory stator is unrectified AC current. However, the battery stator is rectified from AC to DC by a set of diodes. Recommended battery sizes range from 20-amp units in warm climates to 40-amp batteries in colder regions. (See Figures 8-15 and 8-16.)

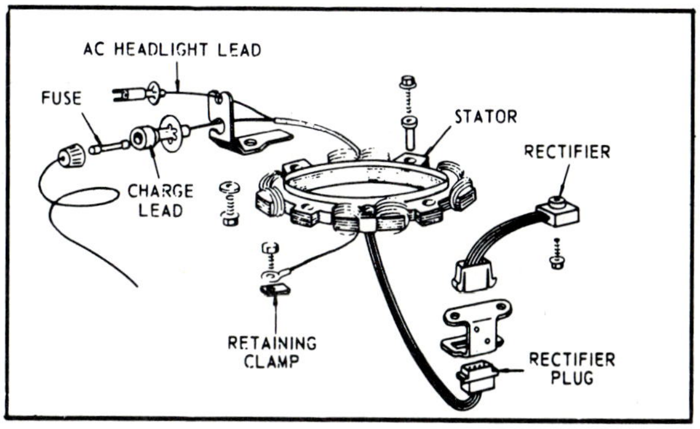

FIGURE 8-16. A dual-circuit alternator assembly (Courtesy of Briggs & Stratton Corporation).

9

LUBRICATING SYSTEMS

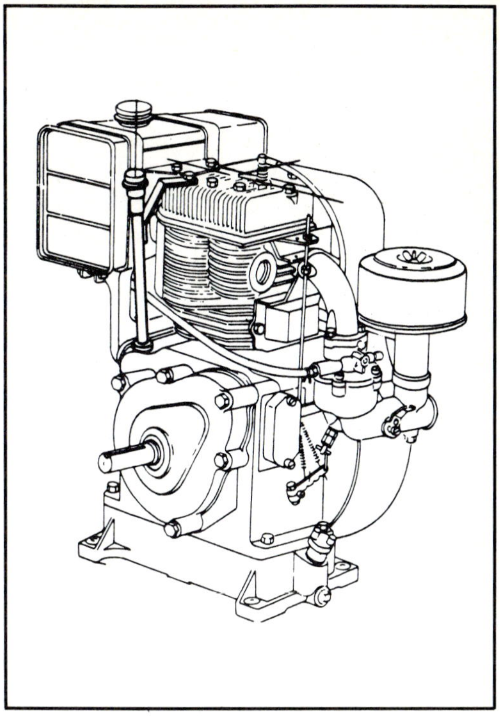

GENERAL

Many small engines receive notoriously poor treatment at the hands of their owners. The care given the lubricating system may be especially bad: infrequent oil changes, using the wrong type or grade of oil, etc. However, lubrication is just as important for small engines as it is for any other type of internal combustion engine. It may even be more important considering the hostile environment in which many small engines operate.

This chapter reviews small engine lubrication, including: basic lubrication concepts, various oil types and simple maintenance procedures, the lubrication system components found in small four-cycle engines and the lubricating procedures for small, two-cycle engines. Pay attention to this information. It could save you time and money.

LUBRICATION FUNDAMENTALS

Friction

Friction can be the enemy or ally of mechanical systems; it depends on their function. Brakes, for instance, couldn't operate unless there was resistance to sliding contact between the brake shoes and the drums. Friction absorbs the energy of the vehicle's motion, changing it into low-grade heat, dissipated by the conductive surfaces of the brake assembly.

However, friction is the decided enemy of most engine parts that slide against one another. Even the smoothest surfaces contain microscopic irregularities that tend to catch and cause resistance to motion.

The way to reduce destructive friction is to prevent parts from touching. That is the primary job of lubricating oils.

Action of Lubricating Oils

Lubricating oils form a wedge or barrier between two moving surfaces. In a sense, the oil becomes a two-layer "sandwich." One oil layer or half of the sandwich adheres to one moving surface, and the other oil layer adheres to the other surface. Sliding contact is restricted to the motion between the two layers of oil—with the adjacent molecules of oil acting like tiny ball bearings. Although friction still exists due to the cohesion between the oil molecules, it is much less than the friction that would result from dry surfaces rubbing together.

The lubricating action between a shaft and a bore is similar, except that the wedging action is emphasized. Oil that is constantly introduced into the space between a bore and rotating shaft (as in a crankshaft and journal assembly) is shaped into a wedge by the weight and motion of the shaft. This oil wedge lifts the shaft into a slightly eccentric position with respect to the bore and separates the shaft from the bore.

Other Functions of Lubricating Oils

Besides reducing friction, lubricating oils also perform other functions. They keep the engine clear by washing away debris resulting from wear. Oils also help cool the engine by absorbing heat from the combustion process.

OIL GRADES AND SIMPLE MAINTENANCE

SAE Classifications

The Society of Automotive Engineers provides the most common designations for identifying different kinds of oils used in internal combustion engines: SAE 10, 10W, 20, 20W, 30, 40, 10W-40, etc. These numbers indicate the relative viscosities of oil, SAE 10 being the least viscous and SAE 40 being the most viscous. The multigrade oils (e.g., 10W-30) retain the properties of the lower-numbered oil at lower temperatures and the properties of the higher-number oil at increased temperatures.

Although the SAE system is generally used, there are some common misconceptions about what the terms mean. For instance, the W suffix after lower-grade oils does not mean weight. It simply means that the oil has been designed especially for winter use. Nor do the SAE numbers correspond to proportional differences in viscosity; SAE 40 oil is not necessarily twice as thick as SAE 20 oil.

Recommended Oils

Most small engine manufacturers recommend high-quality oils with an American Petroleum Institute (API) rating of MS (for multiple service). The API also suggests using detergent oils to help keep the engines clean and to retard the formation of gum and varnish deposits.

The particular SAE grade of oil employed depends on the temperature range to which the engine will be exposed. The chart appearing in Figure 9-1 shows the grades of oil used for various temperatures.

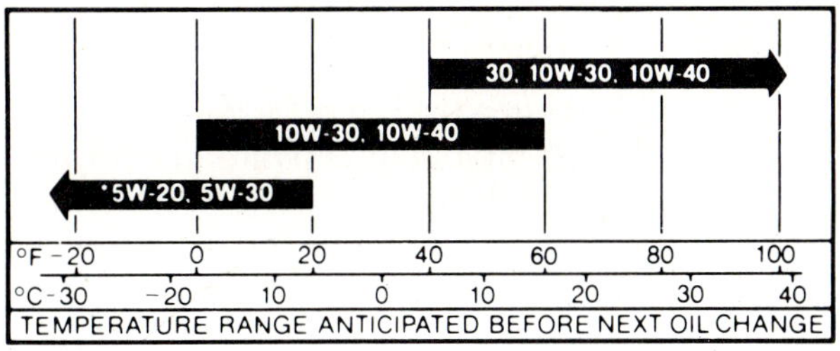

FIGURE 9-1. Recommended SAE viscosity grades (Courtesy of Briggs & Stratton Corporation).

Oil Changes

The oil change interval depends on the manufacturer's recommendations. One manufacturer, for instance, specifies that the oil be changed after the first five hours of operation and every 25 hours thereafter. Regardless of the change interval specified, the oil should be changed more often when the engine is operated in extremely dirty or dusty conditions.

Checking the Oil

It is also necessary to check the level of the oil in the crankcase on a periodic basis. Two basic methods are provided to gauge the oil level: the oil fill plug and the extended oil fill. Both use the same opening for checking the oil and for adding additional oil to the crankcase.

OIL FILL PLUG

This plug, which may be fabricated from plastic or metal, is located on the side of the crankcase. When the plug is removed, the operator simply looks at the exposed oil surface to determine the level. Usually the side of the fill/check hole contains a vertical slot, something like the slot in the water fill hole used in some storage batteries. Most manufacturers recommend that the oil level come up to the top of the slot. (See Figure 9-2.)

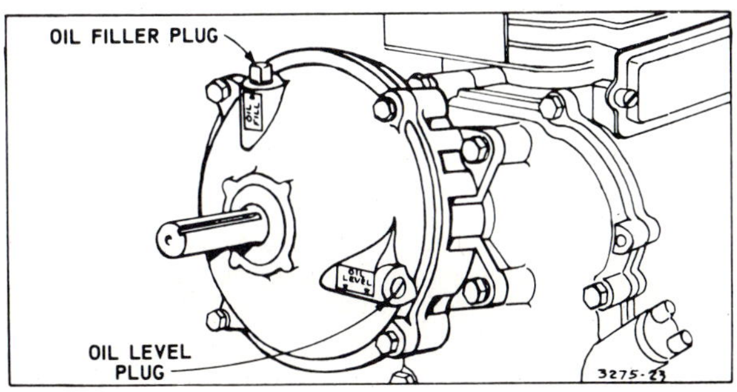

FIGURE 9-2. Oil-level check position (Courtesy of Briggs & Stratton Corporation).

EXTENDED OIL FILL

Sometimes it is not practical to employ a plug to check the oil. Instead, an oil fill pipe is attached to the crankcase. A dipstick, usually formed integrally with the filler cap, is used to check the oil level.

Special care should be taken when installing an extended oil fill pipe and dipstick. Be sure that the "O" ring seal around the bottom of the pipe is properly positioned and that the dipstick is sealed. An air leak can result in a loss of crankcase vacuum, which, in turn, can have a detrimental effect on the lubrication process. (See Figures 9-3 and 9-4.)

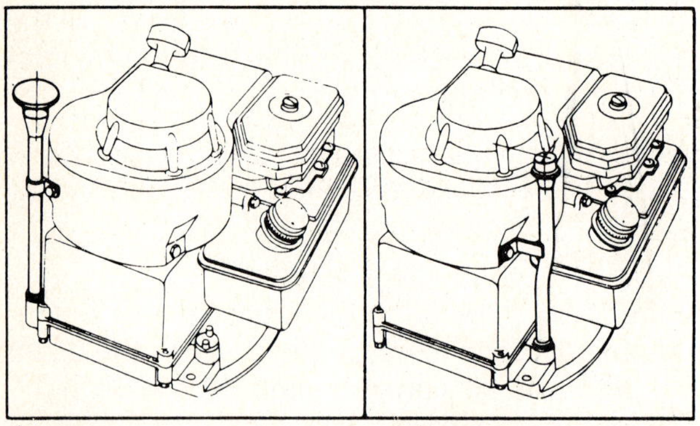

FIGURE 9-3. Extended oil fill tubes for vertical-shaft engines (Courtesy of Briggs & Stratton Corporation).

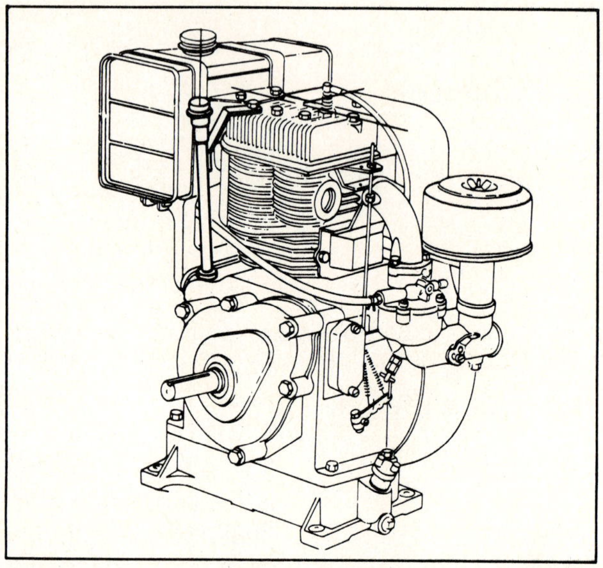

FIGURE 9-4. Extended oil fill tubes for horizontal-shaft engines (Courtesy of Briggs & Stratton Corporation).

Draining the Oil

It is not practical in most instances to drain the crankcase by picking the engine up and pouring the oil out of the fill hole. Consequently, a drain plug is provided on the lower side of the crankcase. The drain plug may be located underneath the deck on vertical shaft engines and in the base of the crankcase on horizontal shaft engines. Some applications attach a small pipe to the drain plug to make it easier to use. (See Figure 9-5.)

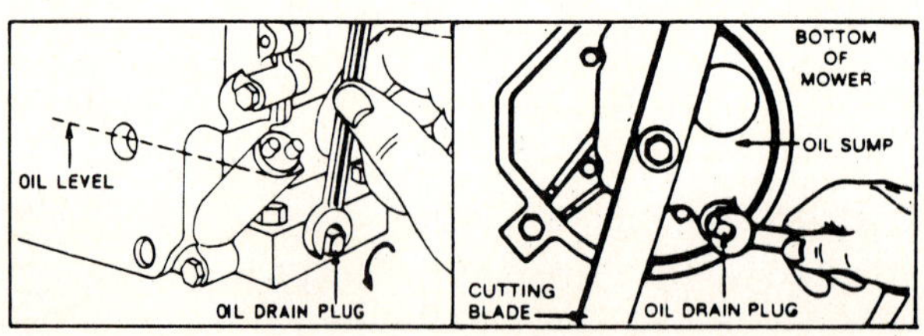

FIGURE 9-5. Oil level and drain plug locations (Courtesy of Briggs & Stratton Corporation).

CRANKCASE BREATHER VALVE

Before getting into the actual operation of four-cycle lubrication system components, we need to briefly examine the crankcase breather valve. Although not directly part of the lubrication system, its operation has an effect on the lubricating process.

The main purpose of the breather valve is to remove gases that have blown by the piston rings into the crankcase. These gases, if allowed to remain in the crankcase, could damage the oil. Also, if no vents are provided, increased pressure in the crankcase (as a result of the buildup of gases) could force oil to leak around the piston rings, oil seals, gaskets, and breaker plunger.

The breather valve in most engines is a spring-loaded fiber disc. The breather may be located inside a valve cover plate, or it may be mounted inside a special housing which leads to a small hole in the crankcase. Gases coming from the crankcase are vented into a pipe connected to the air cleaner. That way, polluting vapors are sent back to the engine, rather than into the atmosphere. (See Figures 9-6, 9-7, and 9-8.)

The breather valve works in this manner: When the piston goes down in the power or exhaust strokes, pressure builds up in the crankcase. The pressure pushes against the breather valve spring, causing the valve to open and vapors to flow through the vent tube to the air cleaner. The vapors are drawn from the air cleaner into the carburetor, where they are mixed with air and fuel, to be subsequently burned inside the combustion chamber.

When the piston goes back up the cylinder, the situation is reversed. A partial vacuum is created in the crankcase. Since the pressure inside the crankcase is now less than the pressure in the vent tube, the valve shuts. The partial vacuum helps to seal off the crankcase and prevent any oil leaks from the areas mentioned before.

LUBRICATING SYSTEMS

There are two main types of lubricating systems employed by small, four-cycle engines—the splash system and the pressurized system.

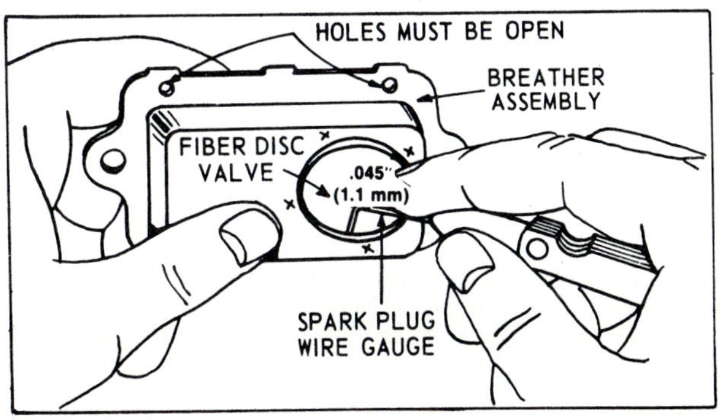

FIGURE 9-6. Checking a valve cover breather (Courtesy of Briggs & Stratton Corporation).

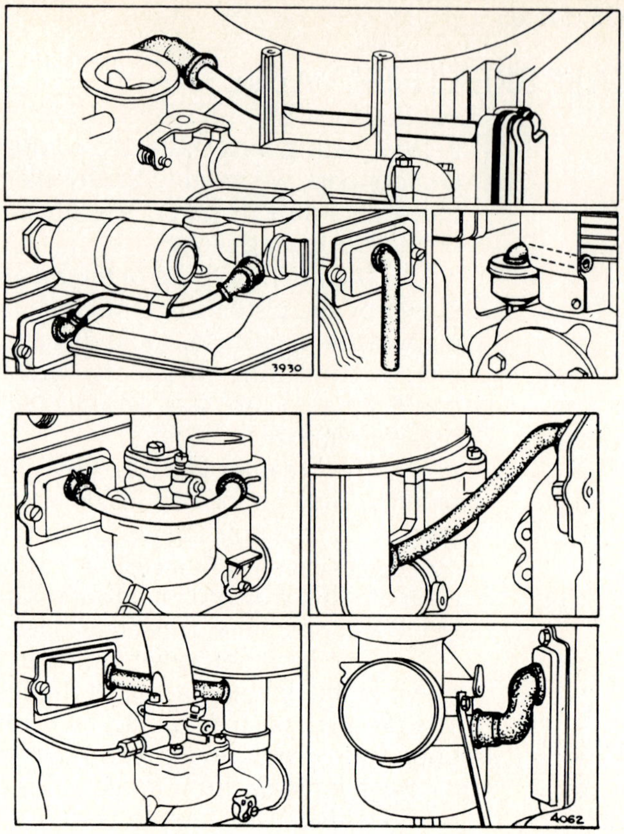

FIGURE 9-7. Breather assemblies (Courtesy of Briggs & Stratton Corporation).

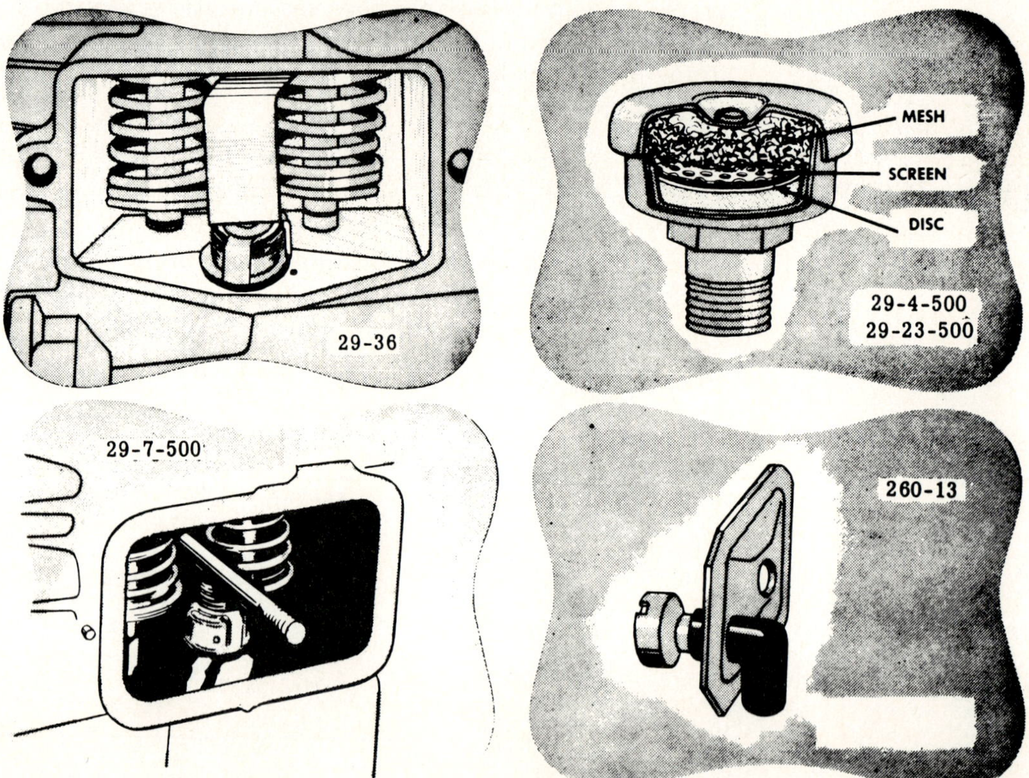

FIGURE 9-8. Valve chamber breathers (Courtesy of Clinton Engines Corporation).

Splash System

The term "splash" effectively describes the operation of this type of lubrication system. A scoop or other similar device is attached to one of the engine's moving components. After periodically dipping into the oil reservoir, the scoop splashes oil on the inside of the engine. (See Figure 9-9.)

The splashing method employed by a given engine depends on the position of the crankshaft and the resulting location of the crankcase oil reservoir. Horizontal crankshaft engines usually locate the dipper scoop (sometimes called oil distributor) on the bottom of the connecting rod cap. Every time the cap passes through the oil reservoir, a fresh supply of oil is scooped up for subsequent splashing. (See Figure 9-10.)

Vertical crankshaft engines rivet an oil scoop to the bottom of the camshaft gear. The top main bearings (as well as other moving and stationary engine components) are lubricated by the circular oil spray produced by the rotating scoop. The bottom bearing, which lies beneath the oil level, is lubricated by the agitation of the oil in the reservoir. The scoop itself spends about 80 percent of its travel under the oil surface, allowing for some fluctuation in oil level. (See Figure 9-11.)

A variation of the splash system employed on vertical shaft engines is the oil slinger. It is a plastic gear mounted on a metal bracket and driven by

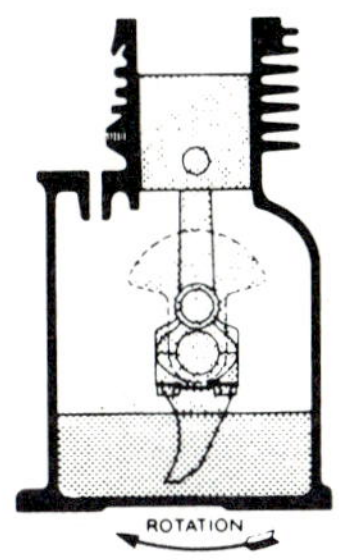

FIGURE 9-9. Principle of horizontal-shaft engine splash lube system (Courtesy of Clinton Engines Corporation).

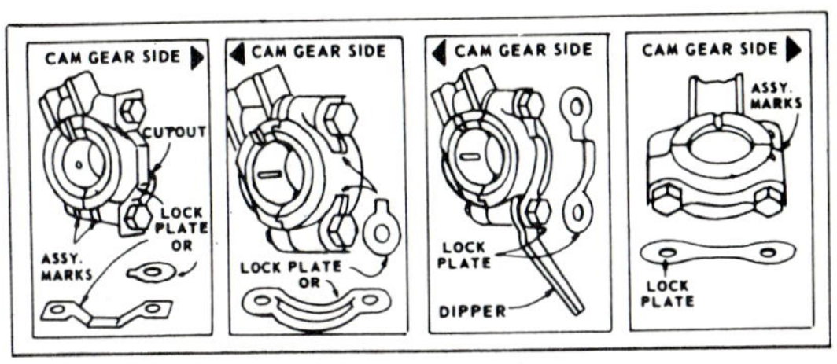

FIGURE 9-10. Splash system dippers for horizontal-shaft engines (Courtesy of Briggs & Stratton Corporation).

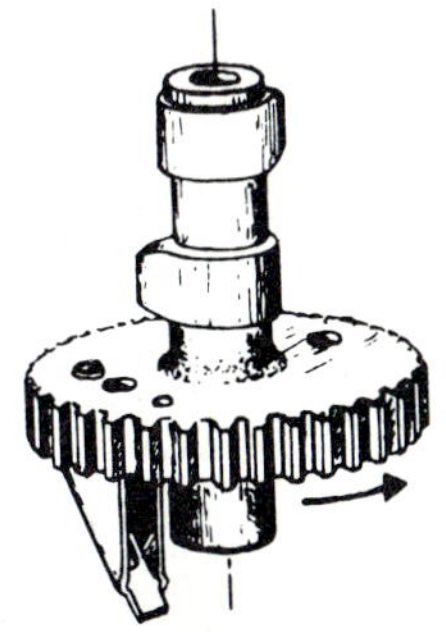

FIGURE 9-11. Oil scoop splash system for vertical-shaft engines (Courtesy of Clinton Engines Corporation).

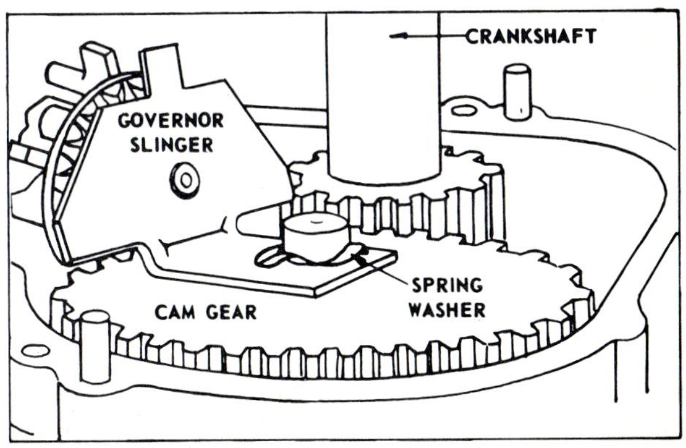

FIGURE 9-12. Gear-slinger splash lube system for vertical-shaft engines (Courtesy of Briggs & Stratton Corporation).

the camshaft gear. Rotating partially beneath the oil surface, the slinger "slings" oil over the inside of the engine. (See Figure 9-12.)

However, just throwing oil over the interior of the engine doesn't necessarily ensure complete lubrication of the most critical rotating components. Therefore, all systems provide oil access holes and/or chamfers (cutout regions) for the main bearings and connecting rod bearings. (See Figure 9-13.)

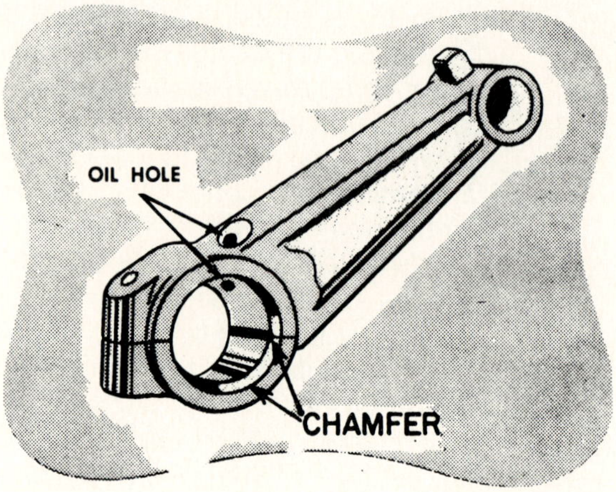

FIGURE 9-13. Oil hole and chamfer on connecting rod (Courtesy of Clinton Engines Corporation).

Pressurized Systems

Splash-type lubrication is not suitable for some small engines. Pumps are required to force the oil under pressure to critical rotating components or wear areas. Oil pumps are usually required for high-performance, high-speed small engines. Pumps are also employed in most multicylinder small engines and in some hand-held four-cycle engines that are operated in various positions.

The oil in most pumps is circulated by a pair of intermeshing gears driven by one of the engine's rotating components. For example, one manufacturer of small two-cylinder engines designs the pump to operate from the crankshaft gear (the gear that turns the camshaft). Oil is supplied to the inlet side of the pump by a pickup screen located beneath the oil level in the oil sump (bottom of the crankcase). The screen filters out large particles and debris. The oil leaves the outlet side of the pump at 25–35 psi (pounds per square inch) pressure. From the pump, the oil proceeds along a series of passageways, called galleries, and through holes drilled in the bearings and shafts. The system also includes a pressure relief valve. It makes sure that the oil pressure and volume remain relatively constant. (See Figures 9-14 and 9-15.)

Combination lubrication systems are used for some applications of small, vertical-shaft, single-cylinder engines. Pressurized oil flow is provided by a gear pump operated by a drive pin located in the hub of the camshaft. Oil from the pump outlet is forced through lines to the top of the engine. The lines are connected with drilled passages leading to the top

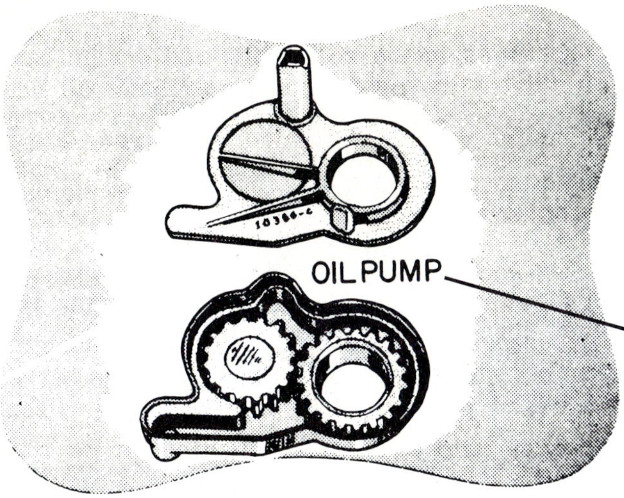

FIGURE 9-14. Oil pump (Courtesy of Clinton Engines Corporation).

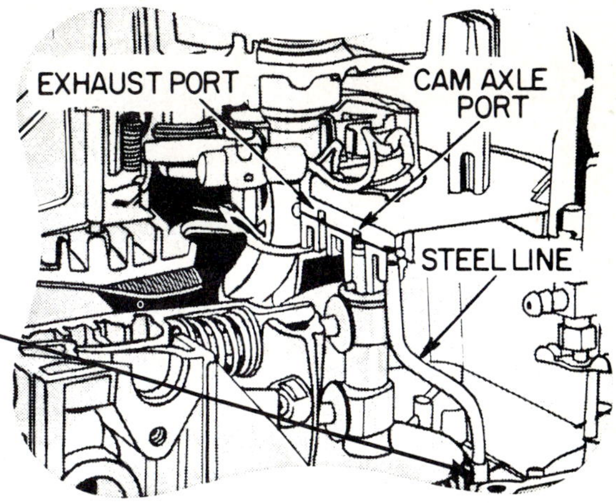

FIGURE 9-15. Oil lines and passages (Courtesy of Clinton Engines Corporation).

main bearings. The connecting rod cap is lubricated by oil squirting from a hole drilled in the main oil line coming from the pump. A positioning notch located outside the hole ensures that the oil flows into a cup or hole in the connecting rod cap. (Naturally, it is very important to make sure that the positioning notch is properly lined up when the engine is reassembled. (See Figure 9-16.)

The lower main bearing and the crank and camshaft gears lie below the oil level in the crankcase. The pressure system is not needed for these components, since they receive adequate lubrication just by splashing through the oil reservoir. In addition, the rod and main bearings throughout the engine are chamfered so that the oil will reach all the bearing surfaces. (See Figures 9-17 and 9-18.)

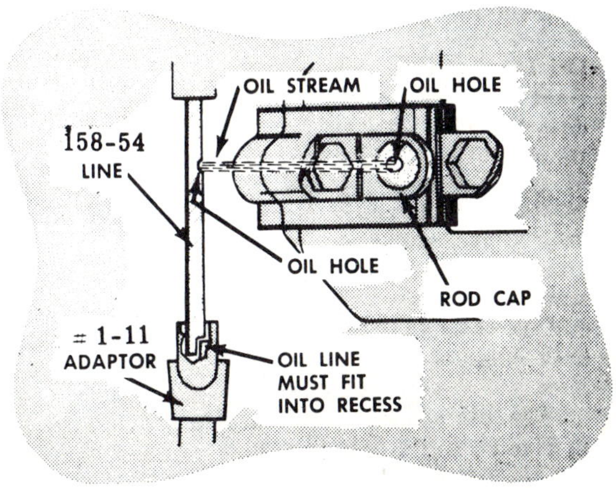

FIGURE 9-16. Alignment of oil line hole (Courtesy of Clinton Engines Corporation).

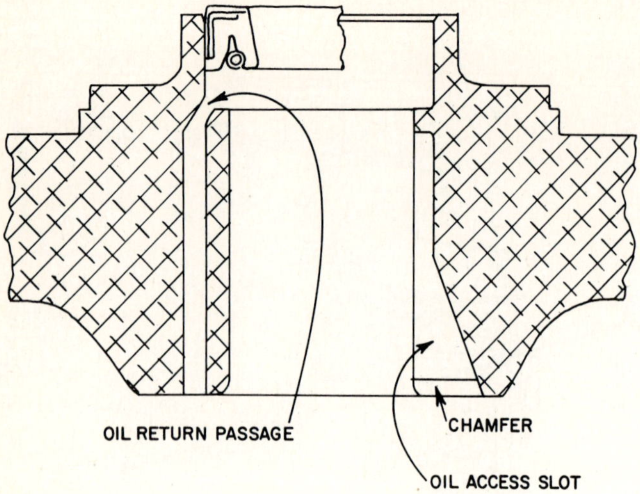

FIGURE 9-17. Chamfered bearings and oil return holes (Courtesy of Clinton Engines Corporation).

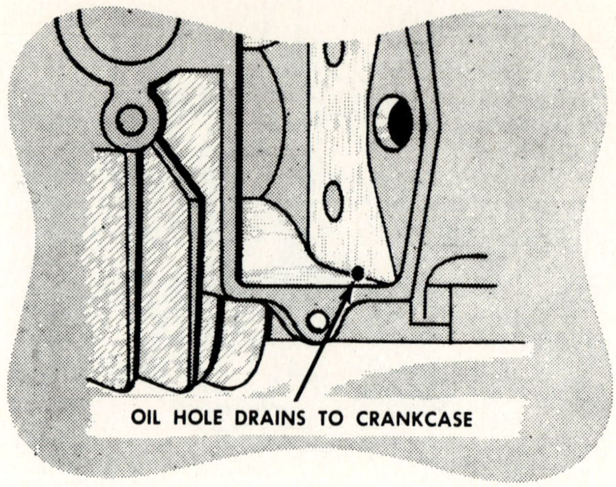

FIGURE 9-18. Valve chamber drain holes (Courtesy of Clinton Engines Corporation).

TWO-CYCLE LUBRICATION

As we have noted in previous chapters, two-cycle engines do not require a separate supply of lubricating oil. Instead, oil is mixed with the gasoline. The rotating components are lubricated when the air/fuel/oil mixture enters the crankcase prior to being transferred to the combustion chamber.

To ensure that bearing surfaces receive the proper lubrication, most rotating components are provided with a number of oil access holes. The rod bearing cap also has a wide "acceptance opening" to make sure that the critical rod bearing surface is adequately exposed to the oily mist. (See Figure 9-19.)

Of course one of the most important functions of two-cycle engine lubrication is to make sure that the gas and fuel are properly mixed. It is also important to ensure that the two-cycle oil specified by the manufacturer is always used. Uninformed experiments with various oil types can have serious consequences.

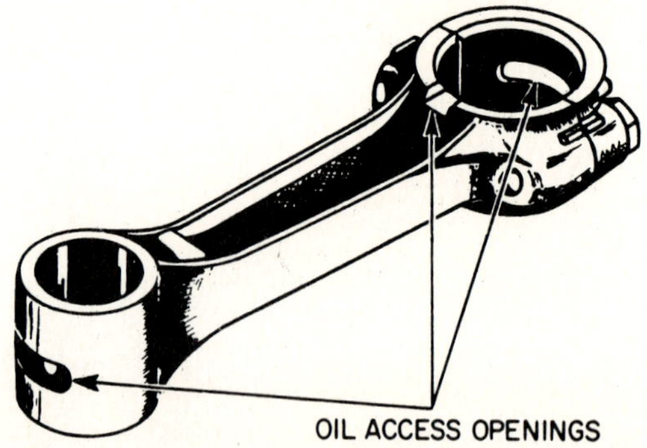

FIGURE 9-19. Access openings in a connecting rod of a two-cycle engine (Courtesy of Clinton Engines Corporation).

10

COOLING AND EXHAUST SYSTEMS

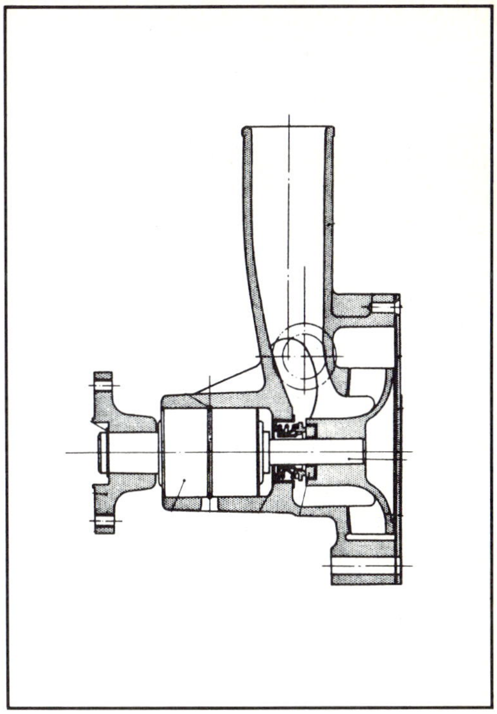

GENERAL

Two important but often ignored aspects of small engine operation are the exhaust and cooling systems. Unless both are in good condition, problems can result. Fortunately, though, the cooling and exhaust systems found on small engines are relatively trouble free.

COOLING SYSTEMS

Heat, as we learned in Chapter 3, is a result of atomic activity. As atoms and molecules bounce around more and more inside a substance, the hotter the material becomes.

Heat in small engines is caused by several conditions. First and foremost is the combustion process itself. The burning air-fuel mixture creates a great deal of heat, which is transferred to the surrounding engine components. Another source of heat is the buildup of pressure inside the combustion chamber. When the molecules of air and fuel are forced together, they tend to vibrate faster, which results in additional heat. A third source of heat is the friction caused by sliding contact between engine parts. The atoms in adjoining surfaces tend to tug and pull at one another, which increases atomic vibration and heat.

At this point, most writers on the subject would simply note that the cooling system is designed to remove heat from the engine. Of course, this is true. However, have you ever wondered why it is necessary to remove heat? What harm would it do to simply let the engine get hot?

Expansion of Solids and Liquids

Excessive heat is bad for an engine because, as the engine becomes hotter, its parts tend to expand. The vibrating molecules actually cause the engine components to increase in size. Some expansion is expected (practically speaking, it cannot be prevented), so engineers design engines with a certain amount of clearance between parts in sliding or rotating contact. However, if engine heat and resultant expansion exceed the design limits, then the components will actually touch. When the condition becomes bad enough, the components might stick together, resulting in a situation known as "seizure."

Note: The factor which engineers must take into account when calculating engine clearances is the coefficient of expansion of the materials being used. This term describes the rate at which a substance expands as its temperature increases. It is interesting to note that various kinds of metals have different coefficients of expansion. This means that you had better not replace an iron piston with one made out of aluminum, since their expansion rates are different. Liquids, on the other hand, have a fairly standard coefficient of expansion (with the exception of water and a few others). Gases have the same coefficient of expansion. This is described by Charles's law, which says in effect that a certain volume of gas will expand exactly $^{1}/_{273}$ of its volume for every 1°C rise in temperature. (See Figures 10-1, 10-2, and 10-3.)

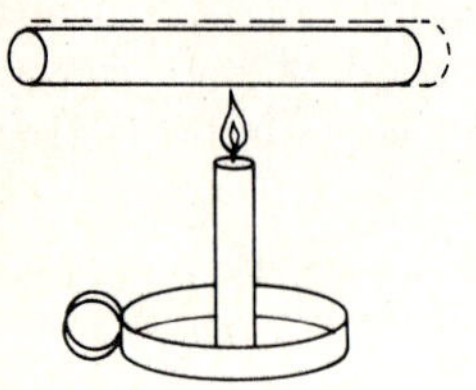

FIGURE 10-1. Expansion of metals.

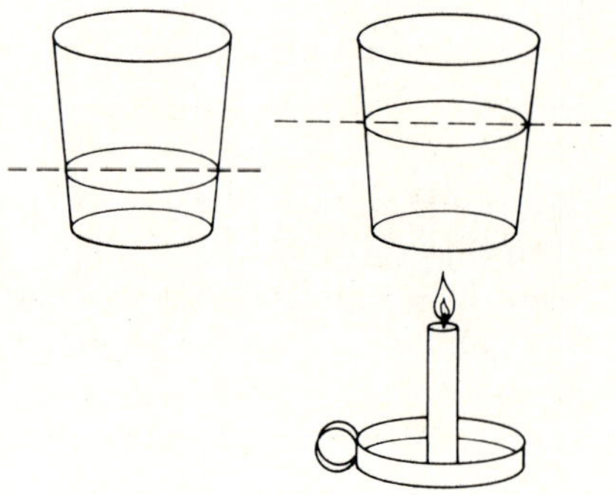

FIGURE 10-2. Expansion of liquids.

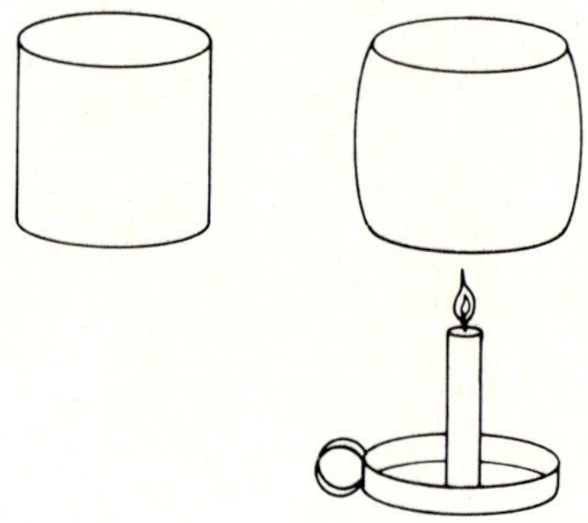

FIGURE 10-3. Expansion of gases.

Removing Engine Heat

Now that we have established that engine heat must be removed to avoid seizure or sticking, the next question is how is the heat removed? There are two basic approaches, the liquid cooling system and the air cooling system.

Liquid Cooling

Liquid cooling systems are used on most larger internal combustion engines and on a few smaller engines. The following paragraphs describe the operation of liquid cooling systems commonly found in automotive-type applications.

Fundamentals of Operation

The cooling system operates by circulating a coolant liquid under pressure through the engine and radiator. Heat is conducted to the liquid in the engine, then radiated into the atmosphere by the radiator. The temperature of the liquid (and hence, the engine) is maintained within a certain range by a thermostat valve. When the engine gets too hot, the valve directs more coolant to the radiator, where the heat is radiated away. When the engine is too cold (e.g., at startup) the valve causes most of the liquid to circulate through the engine, where very little heat escapes.

Coolant Flow and Main Components

The following paragraphs summarize the flow of coolant through the main components in a typical cooling system.

1. *From water pump.* Although the coolant circulates constantly with no real starting point, for the sake of discussion, we can say that the flow begins at the water pump. Most water pumps are of the impeller type, located at the front of the engine and driven by a belt running off the crankshaft. The center part of the propellerlike impeller is a low-pressure region drawing water from the pump inlet. Once the water enters the pump, centrifugal force throws it toward the tips of the blades and the outlet. (See Figure 10-4.)
2. *Into the water jackets.* From the pump, the coolant flows into the water jackets or cooling channels cast into the block and head. The jackets are designed to bring the liquid into contact with the hottest parts of the engine (see Figure 10-5). The coolant begins its passage through the engine at the lower part of the block. As the liquid moves upward in the engine, it becomes hotter. Heated liquids tend to rise (like hot air balloons), which makes the job of the pump easier.
3. *Into the heater coils.* Some of the heated liquid passes from the water jacket through rubber hoses into the heater coils. Air blowing over these

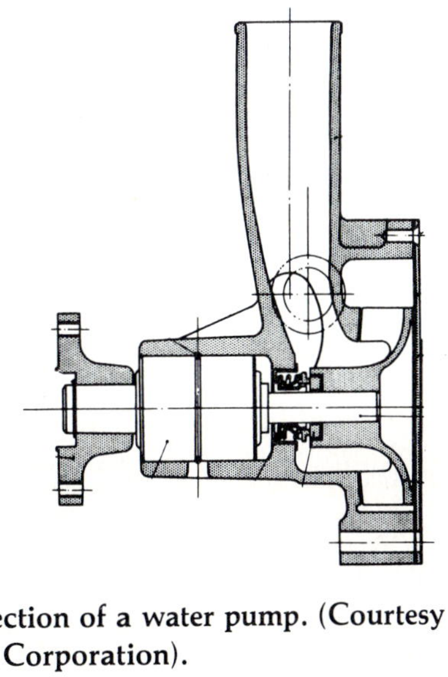

FIGURE 10-4. Cross-section of a water pump. (Courtesy of Nissan Diesel Motors Ltd./Marubeni America Corporation).

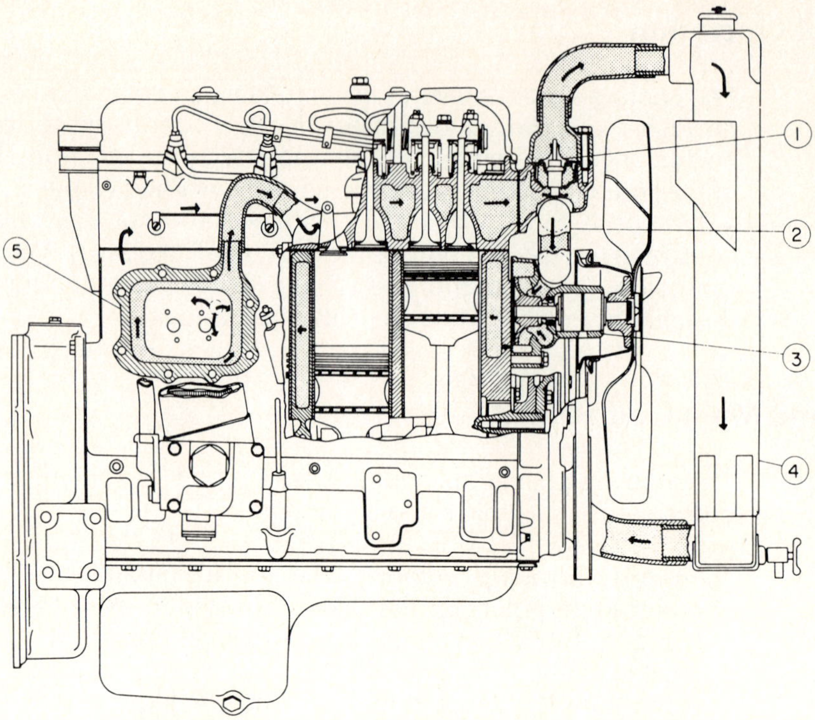

1	Thermostat
2	Bypass hose
3	Water pump
4	Radiator
5	Oil cooler

FIGURE 10-5. Cooling system passages in a Nissan four-cylinder diesel engine (Courtesy of Nissan Diesel Motors Ltd./Marubeni America Corporation).

coils picks up heat and carries it to the passenger compartment. The liquid is returned from the coils back to the cooling system.

4. *To the thermostat valve.* After coolant has circulated through the entire engine, it comes to the thermostat valve (usually located at the top front of the head or block). Most modern engines use a pellet-type thermostat. The pellet is a combination wax and copper sheath. It surrounds an operating rod connected to a valve. When the pellet heats up, it expands and pushes the rod and valve into an open position. When the pellet cools, the opposite action takes place to close the valve.

As noted before, the purpose of the thermostat is to direct coolant to the radiator when the engine is hot and keep it out of the radiator when the engine is cool. The question then arises: How does the coolant get back to the engine when the thermostat blocks the flow to the radiator? (See Figure 10-6.)

5. *To the bypass valve.* The answer in many cases is a bypass valve that allows coolant to flow back to the water pump, where it is directed through the engine again. Several types of bypass valves are used. One type is operated by a spring weaker than the thermostat. Whenever the latter closes, the former opens. Another type of bypass valve is always open, allowing a certain amount of liquid to flow through the engine at all times.
6. *Through the top radiator hose.* When the engine is hot, most, if not all, of the liquid is directed from the thermostat into a hose connecting the top of the engine to the top of the radiator. This hose is often made of molded rubber and it may be (but is not necessarily) reinforced with a wire insert.

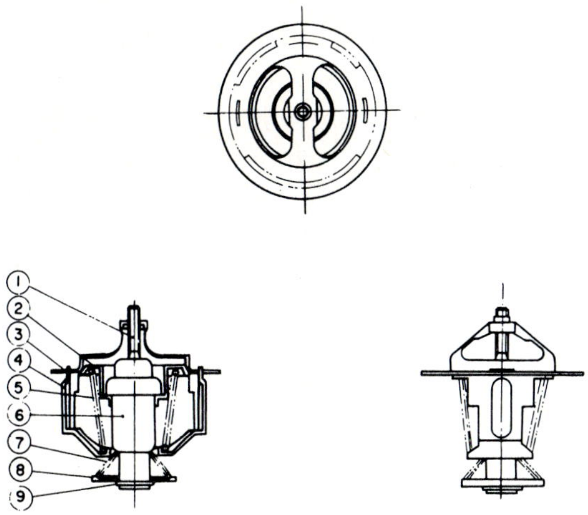

1	Shaft	6	Pellet
2	Valve	7	Spring
3	Sheet	8	Bypass valve
4	Frame	9	Snap ring
5	Spring		

FIGURE 10-6. A thermostat (Courtesy of Nissan Diesel Motors Ltd./Marubeni America Corporation).

7. *Into the radiator.* Once inside the radiator, the hot coolant is forced from the top to bottom through small tubes or passageways. These passageways are connected to a honeycomblike grid of thin metal baffles. Heat is conducted from the passageways to the baffles, which radiate the heat into the air blowing through the intervening spaces. (See Figure 10-6.)

The network of cooling channels and baffles is collectively called the *cooling core*. The reservoir on top where the liquid enters is called the top tank. The bottom reservoir is, naturally, the bottom tank.

The top tank contains the pressure cap and an overflow line to let fluid run off as pressure and temperature go up.

The bottom tank contains a drain valve and connections to the outlet hose.

8. *Through the bottom hose.* After going down through the radiator, the cooled fluid passes into the bottom hose, usually made out of molded rubber and likely to be reinforced by a wire insert. The insert is necessary because the coolant exerts less pressure when it is cool—so little, in fact, that air pressure sometimes tends to collapse the bottom hose.
9. *Back to the water pump.* The bottom hose is often connected to the water pump inlet. At this point, the circuit is complete.

Air-cooled Engines

By far the majority of small engines use an air-cooling system. It is much simpler than liquid cooling; instead of transferring engine heat into a liquid medium and then into the air stream via a radiator, heat is transferred directly into the air. Generally, only a few large engines use air cooling, most of them found in airplanes.

Several features are common to all types of air-cooled engines. For instance, all air-cooled engines have cooling fins cast into the head and block. (See Figures 10-7 and 10-8.) The fins increase the amount of surface area exposed to the air and thereby allow the heat to dissipate faster. Most general-purpose small engines employed in stationary applications also have cooling fans to blow air over the engine. As we have already seen, these fans are constructed integrally with the flywheel. Metal shrouds usually surround the engine to direct the air flow to the hottest areas. (See Figures 10-9 and 10-10.)

The only air-cooled engines that don't use cooling fans are those employed in moving applications. Sufficient air blows over the engine to transfer the heat away.

As a rule, air cooling systems do not require much service. However, operators should make sure that the fins and air flow passages do not become blocked with dirt or debris. Such restrictions could cause "hot spots," which might shorten the life of the engine.

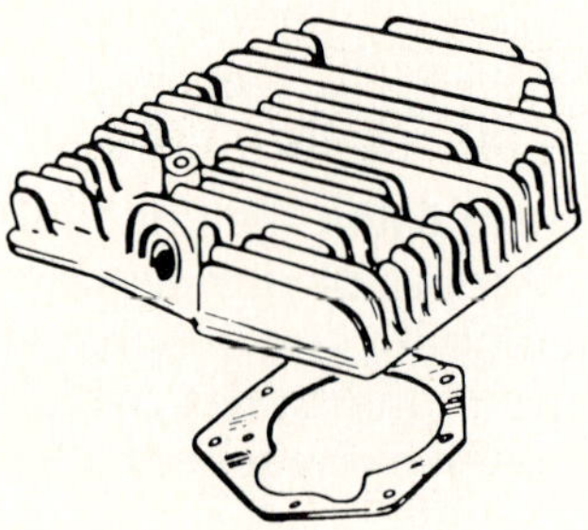

FIGURE 10-7. A cylinder showing air-cooled fins (Courtesy of Onan).

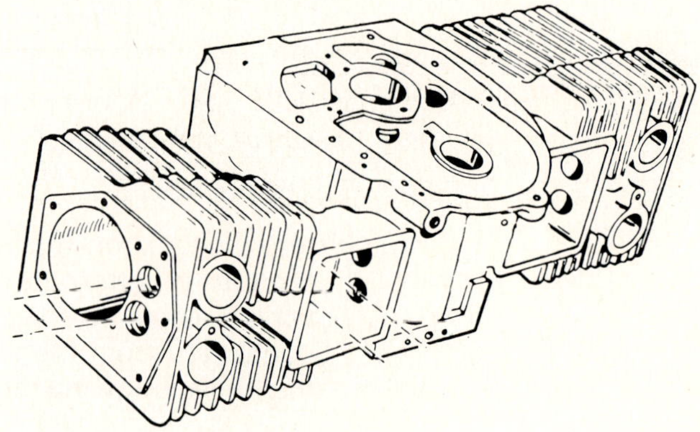

FIGURE 10-8. A two-cylinder air-cooled block showing cooling fins (Courtesy of Onan).

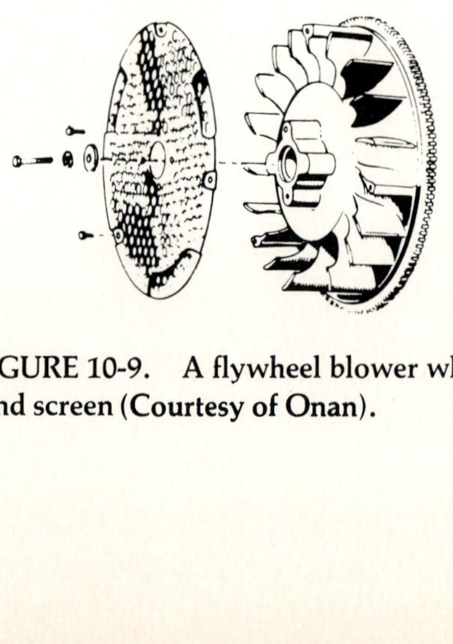

FIGURE 10-9. A flywheel blower wheel and screen (Courtesy of Onan).

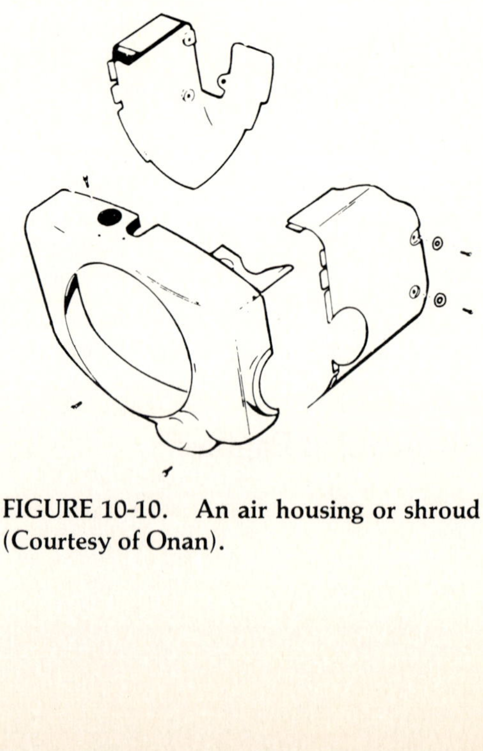

FIGURE 10-10. An air housing or shroud (Courtesy of Onan).

EXHAUST SYSTEMS

Small engine exhaust systems serve several purposes. First of all, the exhaust system provides a passage for exhaust gases to escape from the engine. Second, the exhaust system muffler quietens the often ear-splitting noise produced by many small engines. Third, the muffler helps prevent sparks from going directly into the atmosphere, an important consideration when you are operating a small engine in the woods or anywhere around combustible materials.

Next, we will briefly examine the simple types of exhaust systems found in small single-cylinder and multicylinder engines.

Single-cylinder Exhaust Systems

Single-cylinder exhaust systems usually consist of an exhaust pipe and a muffler. In most cases the muffler is formed integrally with the exhaust pipe. The exhaust gases pass into the short exhaust pipe, which is screwed directly into the exhaust port, and then go from the pipe into the muffler. The muffler contains a series of baffles which cause the exhaust gas to proceed first in one direction and then in another until the sound energy is spent and the noise is reduced to an acceptable level. (See Figures 10-11 and 10-12.)

As simple as this exhaust system seems, engine designers have still managed to come up with variations. For instance, sometimes the exhaust pipe is not screwed into the engine, but is attached by a flange and gasket to the block. In other cases, an exhaust pipe may not be used at all. Instead, a small, flat muffler may be screwed directly to the exhaust port. (See Figure 10-13.)

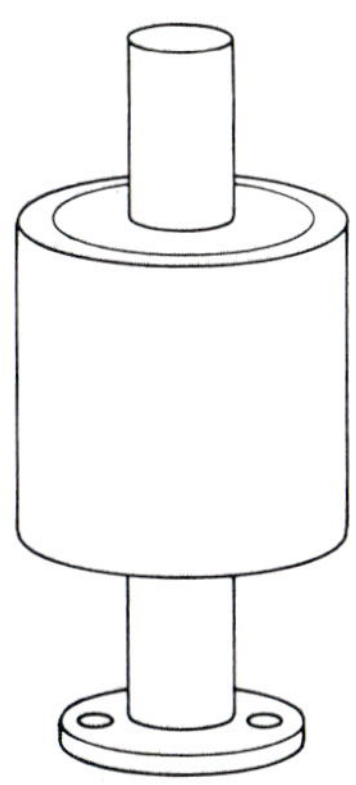

FIGURE 10-11. Typical multicylinder engine muffler.

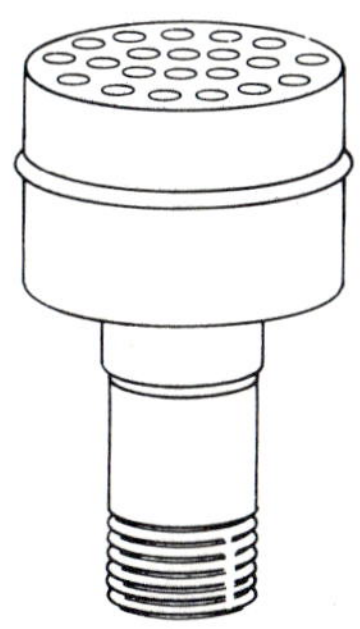

FIGURE 10-12. Typical screw-on-type small-engine muffler.

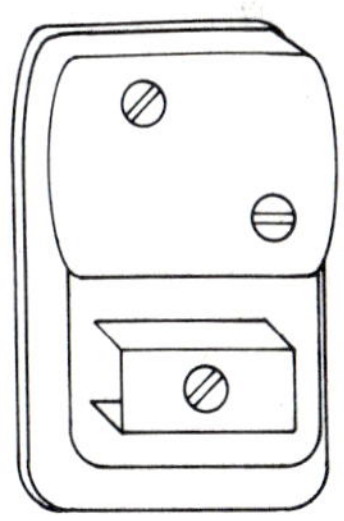

FIGURE 10-13. Small-engine applications of a box-type screw-on muffler.

Multicylinder Exhaust Systems

In addition to an exhaust pipe and a muffler, multicylinder engines also require an exhaust manifold. It is simply a device which allows two or more cylinders to be connected to a single muffler. The exhaust manifold is usually formed by a casting process. The end containing multiple openings is attached by cap screws to the engine's exhaust ports. Sealing between the manifold and the block is provided by an asbestos gasket. The end with a single opening is provided with a flange so that it can be connected to an exhaust pipe, which in turn is connected to a muffler. As in single-cylinder exhaust systems, the muffler and exhaust pipe may be formed as one unit. (See Figures 10-14 and 10-15.)

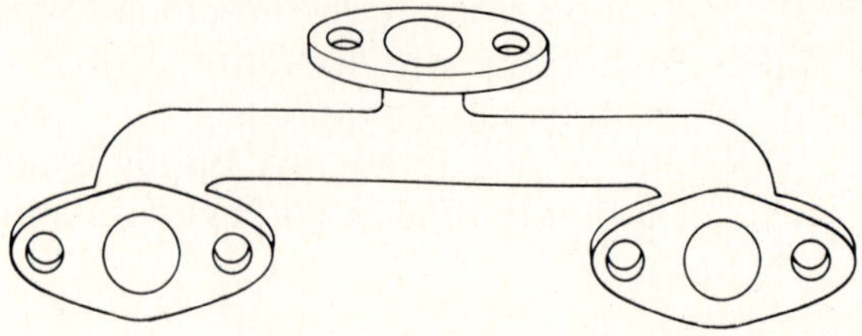

FIGURE 10-14. Typical exhaust manifold for a multicylinder engine.

FIGURE 10-15. A view of a two-cylinder air-cooled engine with the shroud removed. The cooling fins and exhaust muffler can be seen. (Courtesy of Onan).

Exhaust Service

There are really only two kinds of exhaust system service. The first obviously is to replace burned-out mufflers or exhaust. Second and perhaps not so obvious is to make sure that the exhaust system does not become clogged or blocked. Restrictions could cause excessive back pressure, which might burn the valves prematurely.

11

VALVE TRAIN OPERATION AND SERVICE

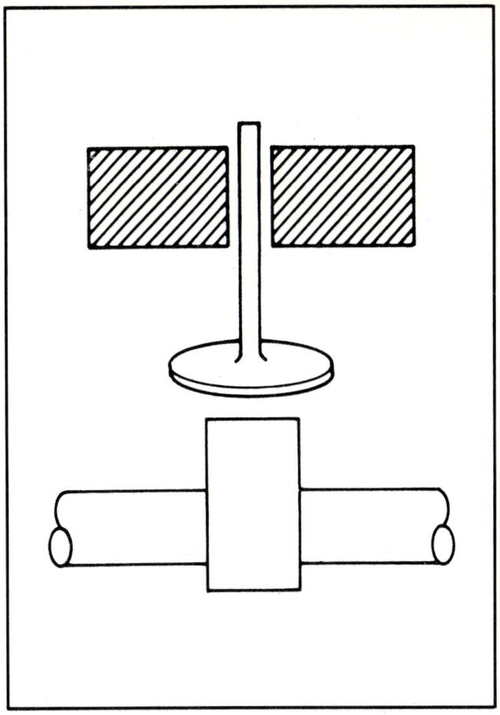

GENERAL

Chapter 4 introduced the major valve train components and how they work together to admit air and fuel into the combustion chamber and remove exhaust gases. The present chapter expands upon the previous discussion, going into more detail about the operation of the valve system. The chapter concludes with a brief examination of the major types of valve train service.

Note: This discussion concentrates on the valve systems employed in four-stroke cycle, L-head (flathead) type small engines. Overhead valve and overhead cam systems, which were introduced in Chapter 4, are much less common in general-purpose small engines. And two-stroke cycle engines do not have a valve train as such.

VALVE TRAIN COMPONENTS

The major components in a typical four-stroke, L-head-type valve train include the valves, valve seats, valve guides, valve springs and retainers, valve lifters, and camshafts.

Valves

The valves act as doors to the combustion chamber, one door for the intake port and one door for the exhaust port. Valves, which are constructed from high-quality steel, are shaped somewhat like mushrooms. Each valve, as noted in Figure 11-1, has a head, margin, face, and stem. Intake and exhaust valves appear alike, although the intake valve is usually larger. Also, the exhaust valve may be constructed from a different steel alloy so that it

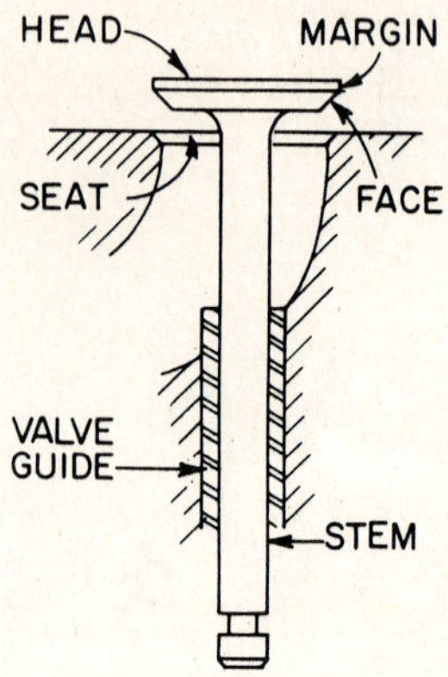

FIGURE 11-1. **Valve part names (Courtesy of Clinton Engines Corporation).**

can withstand heat better. The lower end of the stem of both types of valves is provided with a hole, groove, or some other means of attaching the valve spring retainer.

Valve Seats

Two basic kinds of valve seats are used. Some are machined steel rings pressed into the top surface of the valve ports. Others are cast into the block and machined to final tolerance. This latter kind of valve seat is known as the integral type.

If the valves are the doors to the combustion chamber, then the valve seats are the door jambs or frames. In order for the doors (valves) to close properly, they must fit well within the frames (seats). One way of ensuring a proper fit is to machine the valve face and seat at the same angle. For instance, if the seat is ground at a 30-deg angle, then the valve would also be ground (or "faced") to 30 deg. That way the valve face and seat would contact uniformly all the way around to provide a complete seal. (See Figure 11-2.)

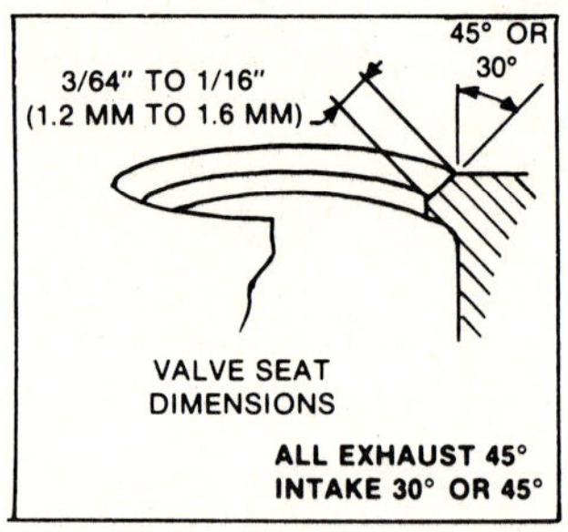

FIGURE 11-2. **Valve seat dimensions (Courtesy of Briggs & Stratton Corporation).**

Valve Guide

Even if the valve face and seat are ground to perfection, the seal will not be correct unless the valve is lowered precisely to the center of the valve seat circle. This is the job of the valve guide. It is located just below the valve port. The guide, which is 1 to 2 in. long, fits around the stem with enough clearance to allow for free up and down movement, but not enough clearance so that the stem can wobble. Many valve guides are pressed into the block and may be replaced when worn. Others are cast integrally with the block. (See Figure 11-3.)

Valve Springs and Retainers

Valves are opened by the action of the cam and lifter (discussed next) and closed by the valve spring. The valve spring is constructed from special steel and is comparatively strong. One end of the valve spring fits against the underside of the valve port. The other end pushes against the valve retainer, which is connected by a hole, groove, or some other means to the end of the valve stem. The stem itself passes through the seat, guide, and coiled valve spring. The action of the spring pushing against the retainer and stem keeps the valve closed until some outside force working in the opposite direction opens the valve. (See Figure 11-4.)

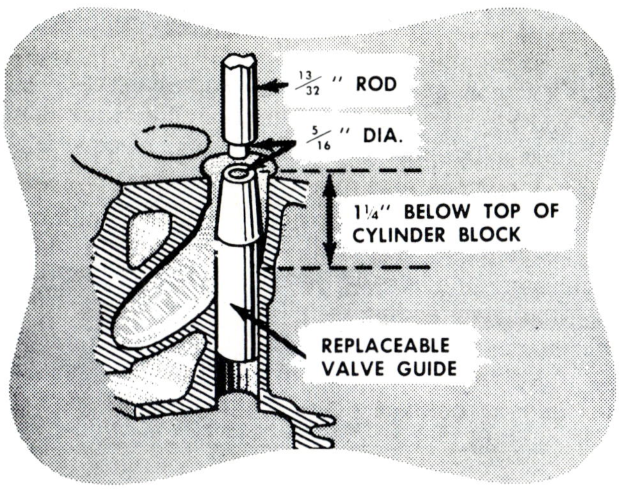

FIGURE 11-3. A replaceable valve guide (Courtesy of Clinton Engines Corporation).

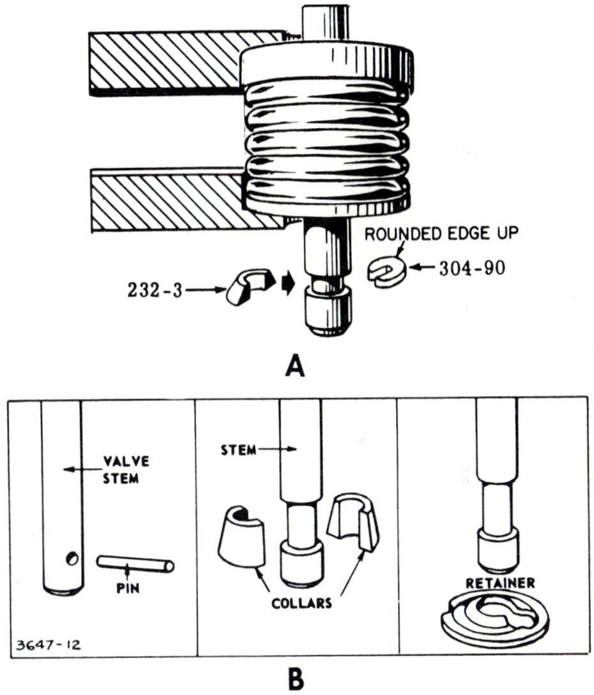

FIGURE 11-4. (A) Valve springs and retainers; (B) three of the more common valve spring retainers (Courtesy of Clinton Engines Corporation and Briggs & Stratton Corporation).

Valve Lifters

The force used to open the valves is transmitted from the camshaft to the end of the valve stem by the valve lifter. The lifter bears a superficial resemblance to the valve it opens. The wide head portion of the lifter rides on top of the cam. The lifter stem passes through a precision hole drilled in the block directly beneath the valve guide hole. The end of the lifter stem engages the end of the valve stem. Therefore, when the cam pushes up on the lifter, the lifter transmits this force to the valve stem to overcome the resistance of the valve spring and open the valve.

Valve lifters, which are made from hardened steel, generally require no service or adjustment. However, the lifters used in some early types of small engines did have a screw for adjusting valve lash (discussed later). See Figure 11-5.

Cams and Camshafts

The cams and camshaft are the "brains" of the valve system. The shape and position of the cams and the rotation of the camshaft determine when the valves open and close and how long they remain open or closed.

The cams are eccentric, or out-of-round, metal circles mounted on the camshaft, one cam for each valve. As the camshaft rotates, the eccentric cam surface rises and falls with respect to the valve lifters. When the cam surface goes up, the lifters are forced up and the valves are opened. When the cam surface falls, the valve springs are then able to push the valves into the closed position.

Cams are designed so that the high point of the cam lobe is preceded by a sloping ramp surface. This causes the valve to be gently lifted up from its seat. The receding ramp (away from the cam high point) may be more steeply angled so that the valve will close faster. However, the ramp will likely flatten out toward the end of the closing period so that the seating operation will not be too abrupt. This is necessary because the valves may have become slightly distorted when exposed to the heat of the combustion chamber.

The camshaft, cams, and cam gear are sometimes formed in one piece from cast iron. However, any number of construction techniques and materials may be used. (See Figure 11-6.)

Valve Timing

Valve timing, or when the valves open and close, is determined not only by the shape and position of the cams, but by the relationship between the gears that turn the camshaft. One of the first concerns is making sure the valves open and close at the proper rate. Since the valves must open and close once every other engine stroke (intake and exhaust), it is necessary that the camshaft rotate at half the speed of the crankshaft. Therefore, the crankshaft driving gear has half as many teeth as the camshaft driven gear.

The valves must also close at exactly the right time, relative to the piston's position, in the intake and exhaust strokes. To ensure that the correct valve timing is maintained, manufacturers provide index marks on the crankshaft and camshaft gears. These marks must be lined up when the engine is reassembled; otherwise, the engine will not run. Consult the manufacturer's manual for the correct valve timing procedure for a particular engine. (See Figure 11-7.)

VALVE TRAIN SERVICE

As in any type of service operation, the best procedure is to first eliminate the simplest possible causes of a problem. Even if you think an engine has

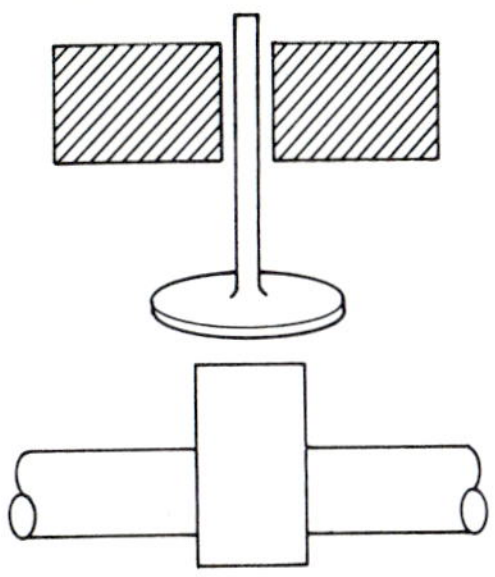

FIGURE 11-5. A typical valve lifter (or tappet) and its relation to a cam.

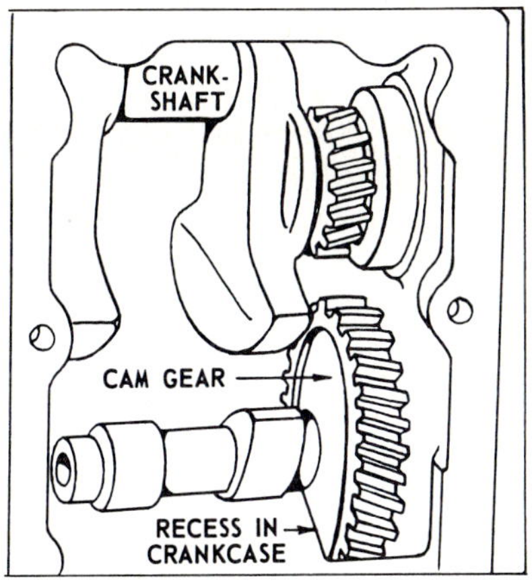

FIGURE 11-6. A typical camshaft and gear (Courtesy of Briggs & Stratton Corporation).

FIGURE 11-7. Valve timing marks on a cam gear and a crankshaft gear (Courtesy of Briggs & Stratton Corporation).

valve train troubles, you should go ahead and check the spark plug, ignition system, and fuel system. Then, if you don't find anything wrong and if a compression check reveals low compression, or if the engine shows no resistance to being turned over by hand, or if it tends to cough back through the carburetor, the valve train may be at fault.

The following paragraphs outline some very general valve train service procedures. As you read this discussion, be aware that no general treatment can possibly cover all the required service procedures for all engines. To save time and ensure a good job, always follow the manufacturer's recommendations for a given engine.

Note: If you aren't familiar with the operation of common hand tools and measuring devices, you may want to skip to Chapters 15 and 16 before proceeding.

Valve and Valve Seat Service

1. First, it is always a good idea to thoroughly clean the exterior of the engine. This makes it easier to disassemble the components; it also helps reveal defects that might otherwise remain hidden by accumulated dirt and debris.
2. Next, remove the air shroud and other sheet metal to gain access to the basic components.
3. Using the correct hand tools, remove the cylinder head. Be careful to note the different lengths of the head screws so that you will know where they go during the reassembly operation.
4. Remove the carbon deposits from the inside of the cylinder head, being careful not to scratch the gasket surface. Thoroughly clean the head and carefully lay it aside for reassembly.
5. Remove the carburetor, fuel lines, and tank (if necessary). This is a good time to clean the carburetor and check the fuel filter and tank.
6. Now, clean and decarbonize the top of the block. Gentle brushing and

scraping will dislodge the carbon deposits from around the valve heads and the piston. Be careful not to gouge the metal as you clean the surface.

7. Remove the exhaust pipe and muffler. Use compressed air to blow loose dirt and carbon from the exposed exhaust port as well as from the head of the engine block and from the intake port. Wear eye protection and be careful not to blow material into the eyes of other workers or onlookers.
8. Using the proper tools, remove the cover from the valve spring compartment. Clean the cover, inspect the breather valve, and lay both aside for reassembly.
9. Select the correct valve spring lifter and remove the valve retainers or keepers. Then remove the valve springs and valves. (See Figure 11-8.)
10. Use a wire brush to clean the valves and valve seats.
11. Replace the valves if they are deeply burned or distorted, or if the margin is less than half that of a new valve. (See Figure 11-9.)
12. Replace or rebuild the seats if they are cracked, burned, or severely distorted. (See Figure 11-10.)
13. In many cases, the valves and seats can be serviced. We won't go into the details of these procedures, since they are usually covered in the manufacturer's manuals. However, we will note that one of three types of servicing procedures are generally employed:
    a. *Hand lapping.* Using an abrasive compound, rotate the valve itself inside the seat to produce the correct face and seat angles. (See Figure 11-11.)
    b. *Power grinding.* A seat grinding machine (resembling a large power drill) is used to establish the correct seat angle. The face angle is ground on a special type of bench grinding machine. Of course, care must be taken in using both of these tools not to cut away too much metal and to make sure that the angles are correct. (See Figure 11-12.)
    c. The proper seat angle can also be obtained by using a hand-operated seat cutter. Again, care must be taken to use the tool properly. (See Figure 11-13.)
14. Reassemble the engine in the opposite order of the disassembly operation.

Checking Stem Clearance Inside the Valve Guide

During the course of putting the engine back together, you should check to make sure that the valve stem fits properly inside the valve guide. As noted before, the stem must move freely inside the guide without wobbling. If the clearance is too great, several courses of action are possible:

1. The valve guide may be replaced. (See Figure 11-14.)
2. A valve with an oversize stem may be used, thereby reducing the clearance.
3. The inside diameter of the existing valve guide may be reduced. This is done by first knurling the inside of the guide. The "knurls" or grooves will project out from the inner wall to make the inside diameter smaller than the diameter of the valve stem. The desired inside diameter is established by using a reaming tool to remove the excess metal. (See Figure 11-15.)

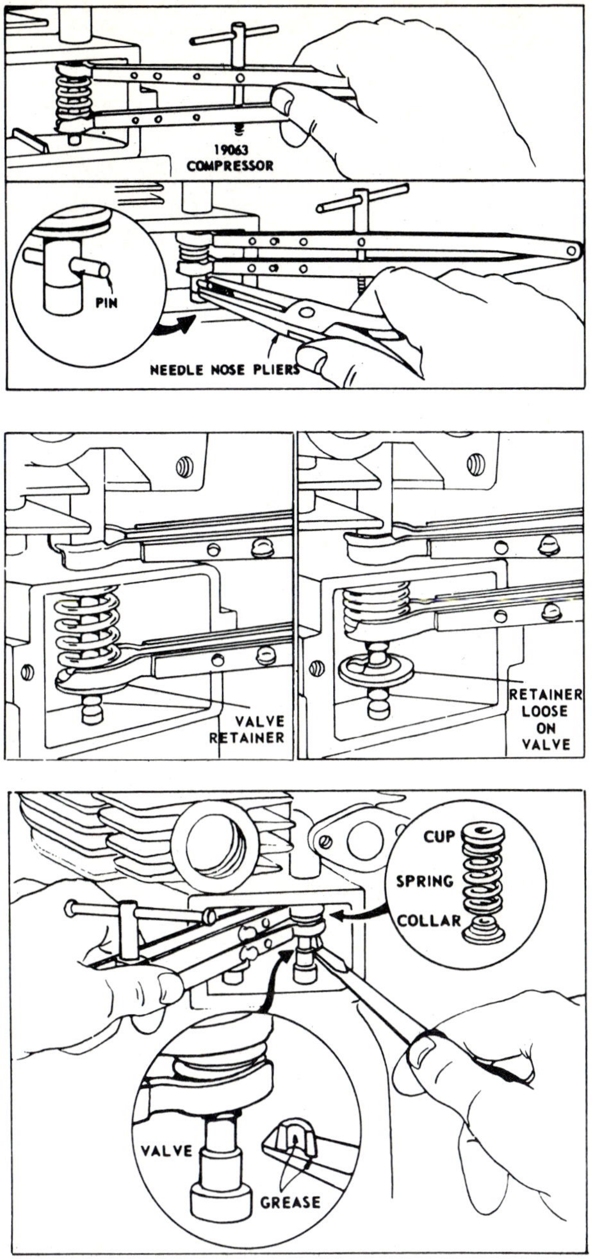

FIGURE 11-8. Removing valve springs and valves (Courtesy of Briggs & Stratton Corporation).

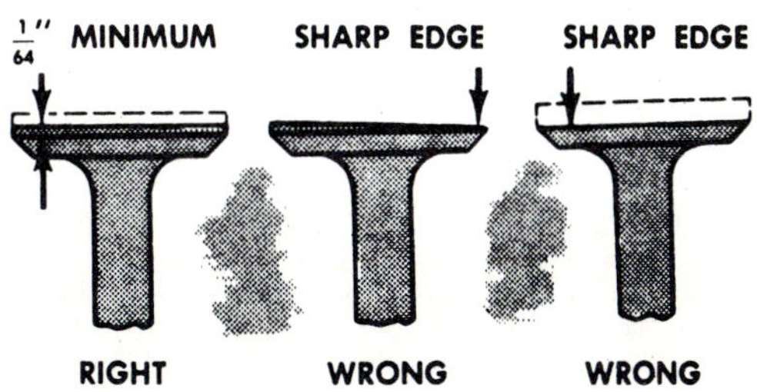

FIGURE 11-9. Inspecting valves, heads, faces, and margins (Courtesy of Clinton Engines Corporation).

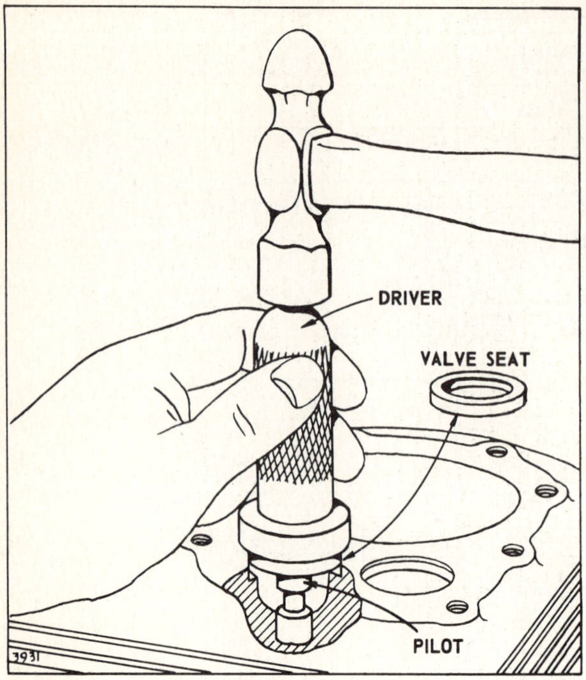

FIGURE 11-10. Installing a valve seat (Courtesy of Briggs & Stratton Corporation).

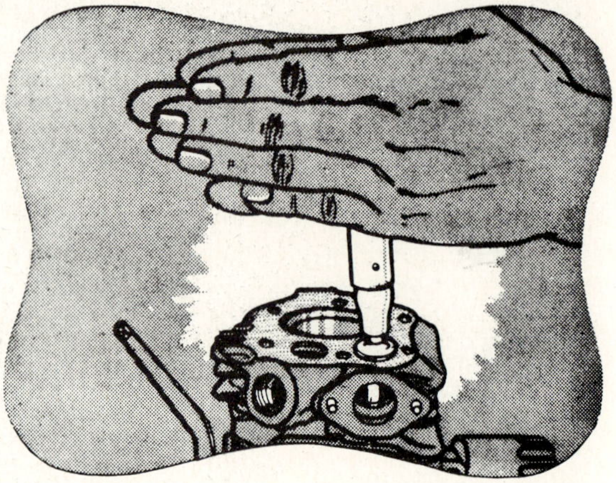

FIGURE 11-11. Hand grinding or lapping valves with abrasive compound (Courtesy of Clinton Engines Corporation).

Checking the Clearance Between the Valve Stem and the Valve Lifter

A manufacturer-specified minimum clearance should exist between the end of the valve stem and the end of the valve lifter. This clearance is sometimes referred to as the valve "lash," or the amount of "play" in the valve train. If there is no play in the valve train, the valve will tend to remain open when it should be closed. This can have several effects. First of all, there will be a loss of compression in the combustion chamber with a corresponding reduction in performance. Second, and perhaps not so obvious, the valve (especially the exhaust valve) will tend to overheat. This is because heat is transferred from the valve to the block when the valve is in contact with the seat. If the valve never touches the seat, then heat will remain in the valve.

Older engines provided a screw to adjust the valve lash. Newer engines do not. Additional clearance is usually obtained by simply grinding some material from the end of the valve stem. This operation usually takes place during a service operation. (See Figure 11-16.) Consult the manufacturer's specifications for specific engines and models.

Completing the Valve Service

The last aspects of valve service really apply to any general engine service: Make sure that all the parts are clean before they go back together. Select the correct gaskets. Use a torque wrench to tighten all the head screws, making sure that you observe the manufacturer's recommended tightening sequence.

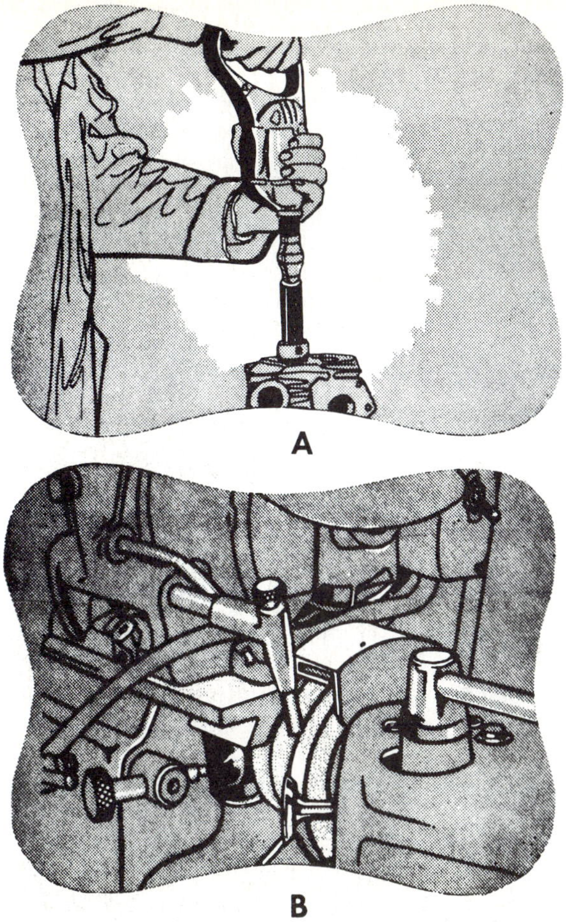

FIGURE 11-12. (A) Grinding a valve seat with a precision seat grinding machine; (B) grinding a valve face using a valve refacing machine (Courtesy of Clinton Engines Corporation).

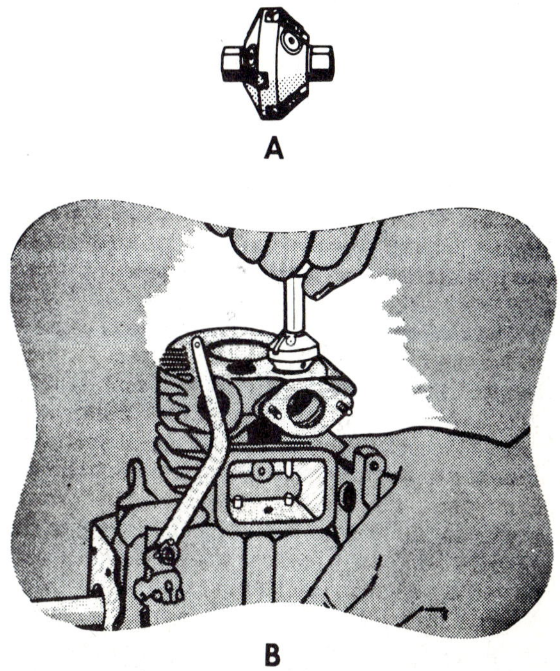

FIGURE 11-13. (A) A valve seat cutter; (B) reconditioning the valve seat using a valve seat cutter (Courtesy of Clinton Engines Corporation).

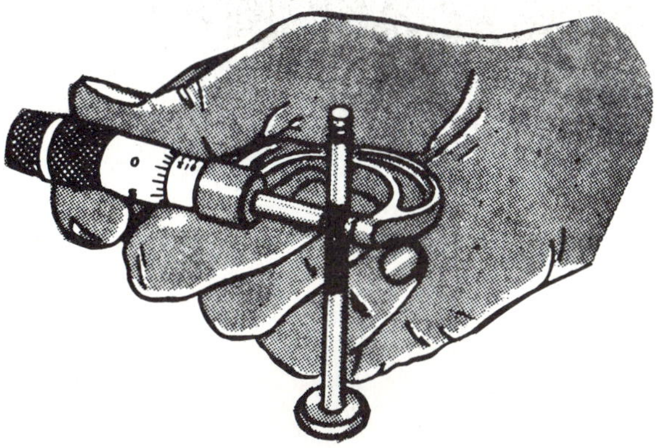

FIGURE 11-14. Checking valve stem wear with a micrometer (Courtesy of Clinton Engines Corporation).

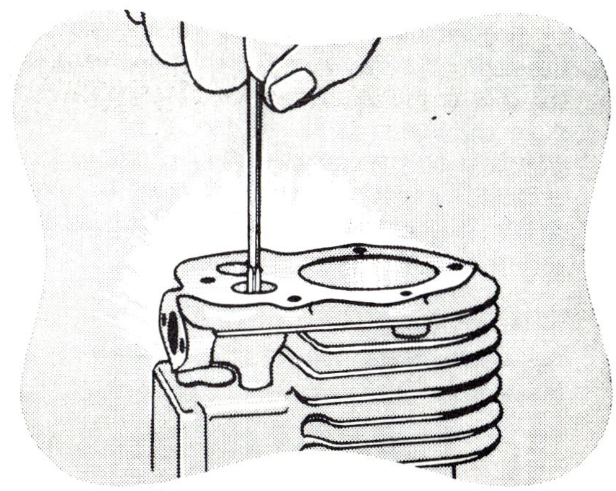

FIGURE 11-15. Reaming knurled valve guides (Courtesy of Clinton Engines Corporation).

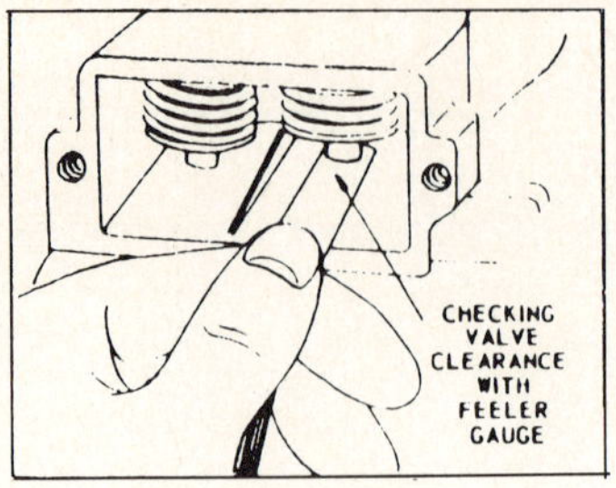

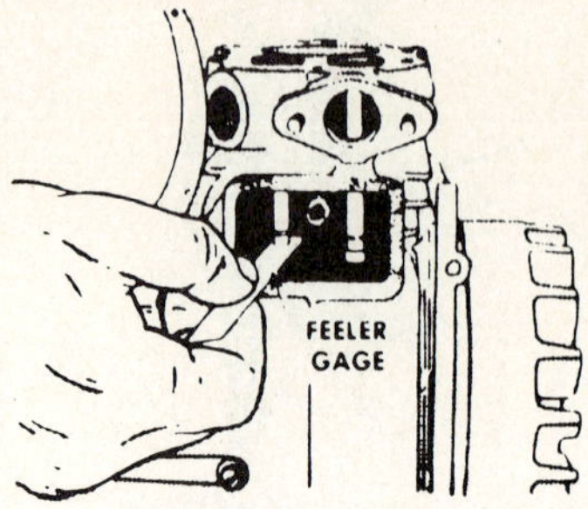

FIGURE 11-16. (A) Checking valve clearance with a feeler gauge; (B) checking valves with a feeler gauge (Courtesy of Briggs & Stratton Corporation and Clinton Engines Corporation).

Camshaft and Lifter Service

Camshaft and lifters cannot be serviced without completely disassembling the crankcase. Normally the camshaft and lifters do not cause problems. If difficulties do arise, the defective units are generally replaced rather than repaired. Follow the manufacturer's directions for a particular engine. (See Figure 11-17.)

TWO-STROKE CYCLE VALVE SYSTEMS

As we noted earlier in this chapter, as well as in previous chapters, two-cycle engines do not have a valve train as such. Port holes in the piston skirt and cylinder wall provide passageways for intake and exhaust gases. The only component that may require service is the flapper-like "reed" valve, which controls the flow of air and fuel from the carburetor to the crankcase. (See Figure 11-18.) However, defective reed valves are not repaired but simply replaced.

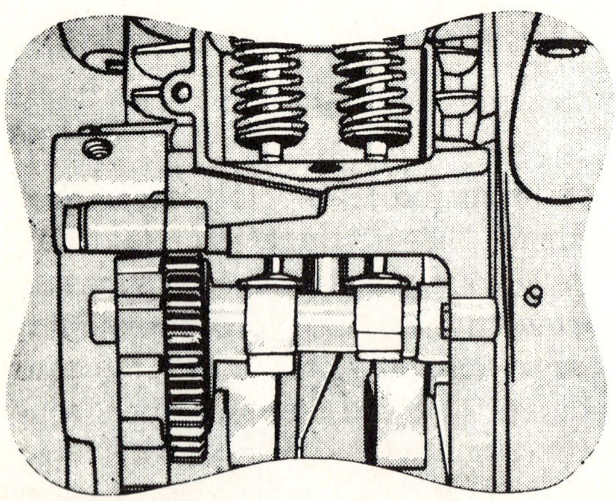

FIGURE 11-17. Cutaway showing cam, tappets, valve stems, and springs (Courtesy of Clinton Engines Corporation).

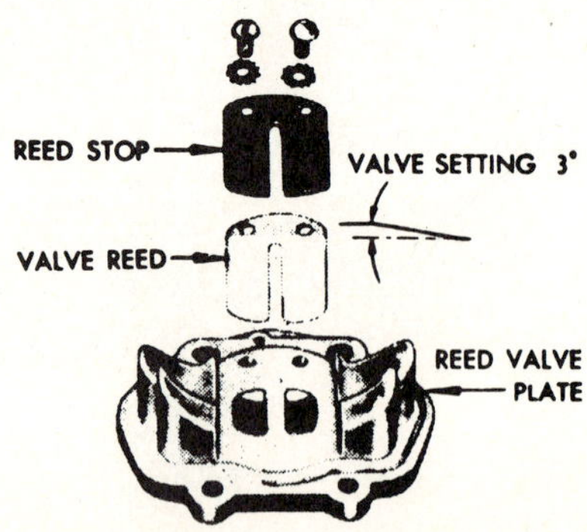

FIGURE 11-18. Reed valve, typical of two-cycle valves (Courtesy of Clinton Engines Corporation).

12

PISTON AND CONNECTING ROD SERVICE

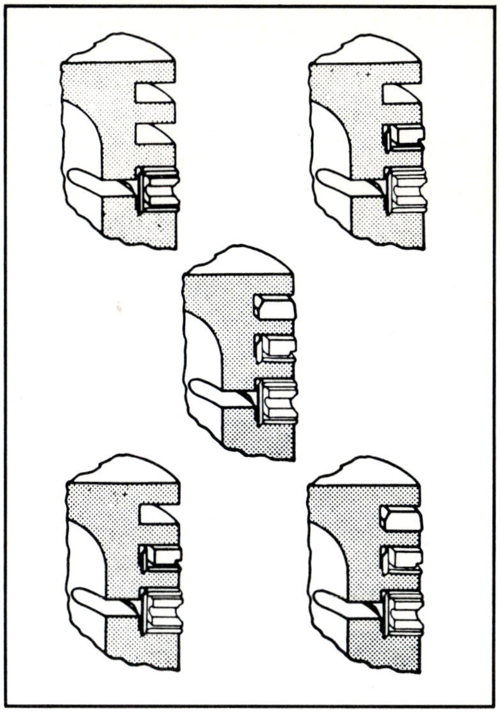

GENERAL

As we learned in Chapter 4, the piston is a movable plug which moves up and down in the cylinder. The piston transmits the force of expanding gas created in the combustion chamber through the piston pin and connecting rod to the rotating crankshaft. (See Figure 12-1.) This chapter expands on the previous discussion of pistons and related components. The chapter concludes with a review of piston and connecting rod service procedures. Before going on, we should note, as we have before, that this book deals with general applications and basic principles. The "real world" contains many more examples than those shown here. Always consult the manufacturer's reference materials before attempting your own service operations.

COMPONENTS

Piston Head

Most small engine pistons are made from cast aluminum. The upper end of the piston is known as the *head*. The lower part, below the bosses (discussed next) is called the *skirt*. The head is usually the thickest part of the piston, since it must withstand the greatest pressures. The head also usually has the smallest diameter. This is so that the piston rings, when they project out from the head, will form the widest diameter. (See Figure 12-2.)

The top of the head may take several shapes, depending on whether the engine is a two- or four-stroke design. Most four-stroke engines employ a flat head piston. Most two-stroke pistons contain some kind of projection or baffle protruding from the top of the head. The tops of some piston heads may also be formed into dome or wedge shapes. These odd shapes force the incoming air and fuel to flow in swirling paths, thereby promoting better mixing. (See Figure 12-3.)

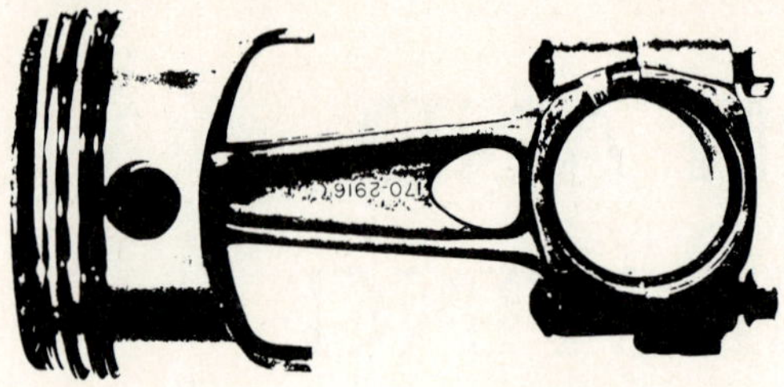

FIGURE 12-1. Typical piston, piston rings, piston pin, and connecting rod (Courtesy of Onan).

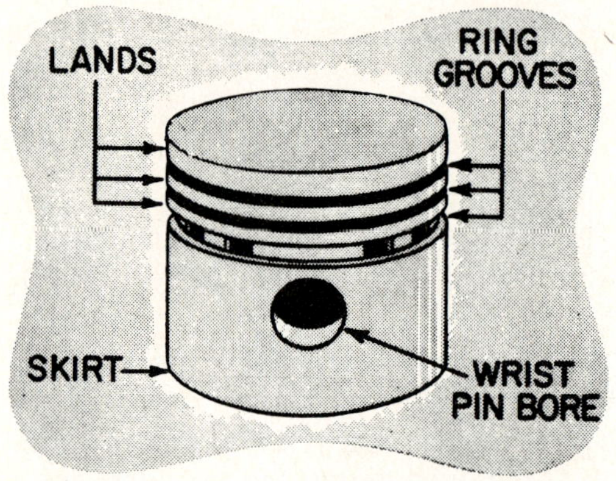

FIGURE 12-2. Piston part names (Courtesy of Clinton Engines Corporation).

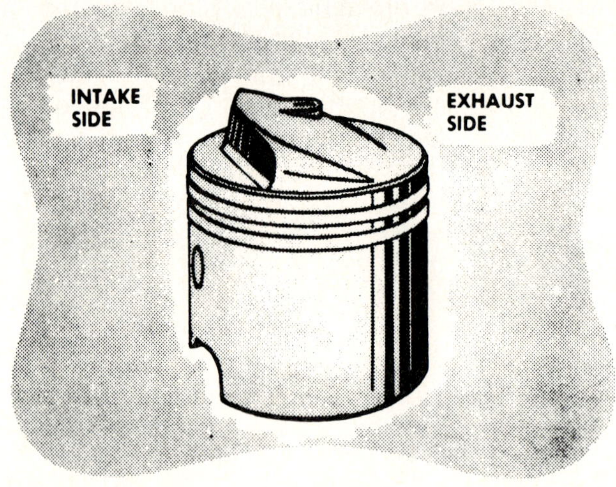

FIGURE 12-3. A piston for a two-stroke cycle engine (Courtesy of Clinton Engines Corporation).

Piston Skirt

The piston skirt guides the piston as it moves up and down in the cylinder. The cross section (end view) of some skirts is circular; others are elliptical or slightly out of round. Elliptical designs, which are commonly referred to as *cam ground pistons,* provide a closer fit with the cylinder bore. This is because all pistons, when they become hot, expand toward the cylinder walls. To avoid seizure, circular cross section pistons must have considerable clearance around the entire skirt. Therefore, when the engine is cold, before any expansion takes place, the piston is likely to "slap" against the cylinder wall. Cam ground pistons avoid this problem by providing a closer fit at right angles to the piston pin (0.003 in. being typical) and a looser fit in the other direction. As the piston warms up, it expands into the area where the cold clearance was greatest to become more round. This design is almost universally used in automotive engines. (See Figure 12-4.)

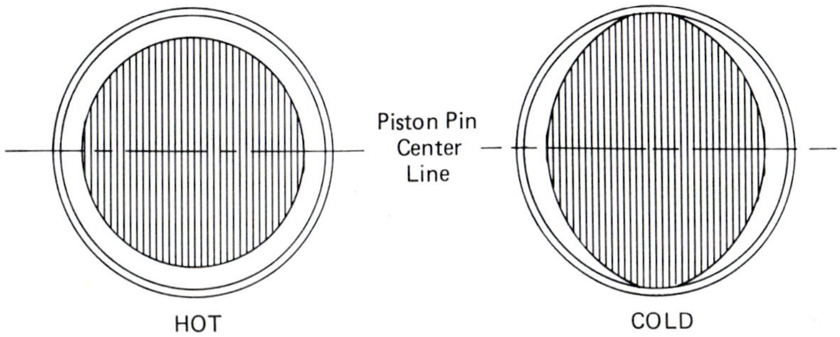

FIGURE 12-4. Principle of cam ground piston design.

Pin Bosses

Pin bosses are the reinforced area where the pin passes through the piston. Such reinforcement is necessary to withstand the forces exerted by the piston on the pin as the combustion gases expand. Piston pin bosses are precisely machined to fit the pin with a very slight clearance. Usually each boss contains a groove to hold a pin retaining spring. In most cases, the piston pin is free to rotate within the bosses. (See Figure 12-5.)

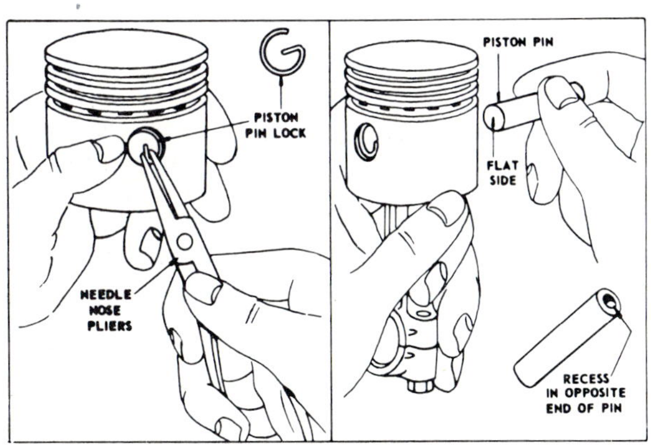

FIGURE 12-5. Piston pins and bosses (Courtesy of Briggs & Stratton Corporation).

Connecting Rod

Connecting rods, or "wrist pins," connect the piston to the crankshaft. Connecting rods may be constructed from iron, steel, aluminum, or some other kind of alloy—although most small engines use aluminum connecting rods. Usually constructed in an "I" beam cross section for strength, the connecting rod has two distinct ends. The small end is reamed to fit around the piston pin. The large end is usually provided with a cap, which is held in position by two screws or bolts. The cap secures the rod to the crankshaft throw.

Connecting Rod Bearings

Many small four-stroke engines do not employ any sort of rod bearings, other than the soft aluminum of the rod itself. Sometimes bronze alloy may be mixed with the aluminum at the larger end of the rod.

Small four-stroke engines subjected to greater loads may be provided with bearing half shells at the larger end of the rod. Common on all automotive applications, these shells contain a steel back and a soft copper alloy bearing surface. One bearing shell is fitted into the end of the rod itself and the other shell into the cap. A location "tang" prevents the bearing from rotating with the crankshaft.

At the upper end of the price and performance range are the roller bearings used on some high-speed two-stroke engines. The roller bearings are mounted in half sets on the large end of the rod to provide an extremely low friction connection.

Piston Rings

Most small four-stroke engines have three rings just below the piston head. The rings, which may be coated to reduce wear and speed "break in," are located in recessed grooves. (See Figure 12-6.) The spaces between the grooves are called "lands." The top two rings are referred to as *compression rings*, while the bottom ring is known as the *oil control ring*.

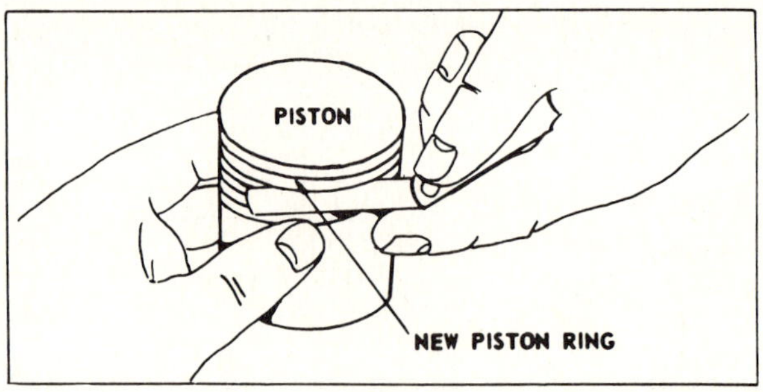

FIGURE 12-6. Checking ring grooves for wear (Courtesy of Briggs & Stratton Corporation).

PISTON, PISTON RING, PISTON PIN, AND CONNECTING ROD SERVICE TIPS

Note: If you are not familiar with the operation of common hand tools and measuring devices, you may want to skip to Chapters 15 and 16 before proceeding.

Disassembly and Cleaning

The first order of business is to disassemble the engine, following the directions given in the manufacturer's reference materials. Next, the components are cleaned. Use recommended cleaning agents and be careful to follow all safety precautions. Avoid flammable materials. After the parts have been cleaned, lay them out in clear view for inspection and subsequent service.

Piston Service

Cast pistons are easily damaged, so be careful when removing them from the block. If a ridge of carbon has built up around the top of the cylinder, use a ridge reamer to remove the deposit before extracting the piston. Several types of ridge reamers are available.

After the piston has been lifted out, carefully remove the rings. Then, using an old broken ring or other tool, scrape the ring grooves clean. This is especially important because any material left in the groove will push out on the ring, causing it to gouge the cylinder wall. (See Figure 12-7.)

While cleaning the piston, inspect for the following conditions:

1. Scraped piston wall
2. Cracked or broken skirt
3. Damaged ring lands
4. Seriously worn or damaged wrist pin, lock ring grooves

If any of these conditions are present, the piston must be discarded. Pistons that pass the visual inspection are then measured by using a micrometer and small hole gauge. The following diameters are checked:

1. *Skirt diameter.* The reading should be made at right angles (90 deg) to the wrist pin bore, near the bottom of the skirt. (See Figure 12-8.)
2. *Land diameter.* This reading should also be made at right angles to the wrist pin bore. Since pistons used in four-stroke engines are usually

FIGURE 12-7. Cleaning ring grooves with ring groove cleaner (Courtesy of Clinton Engines Corporation).

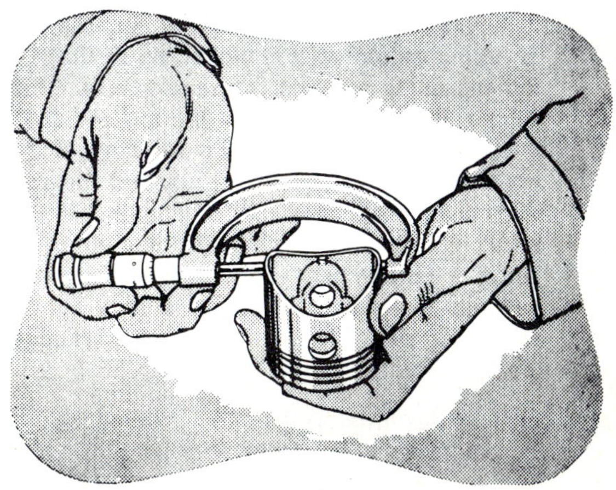

FIGURE 12-8. Measuring the piston skirt (Courtesy of Clinton Engines Corporation).

tapered, you will probably find that the lands at the top are larger than those lower down on the piston. Commonly, a four-stroke piston tapers in about 0.00125 in. per inch of the piston's length. Pistons used in two-stroke engines are normally straight. (See Figure 12-9.)

3. *Wrist pin bore.* Use a small hole gauge and micrometer to check the inside diameter of the wrist pin bore.

These readings are checked against the manufacturer's specifications. If any readings do not fall within the allowed range, the piston should probably be discarded.

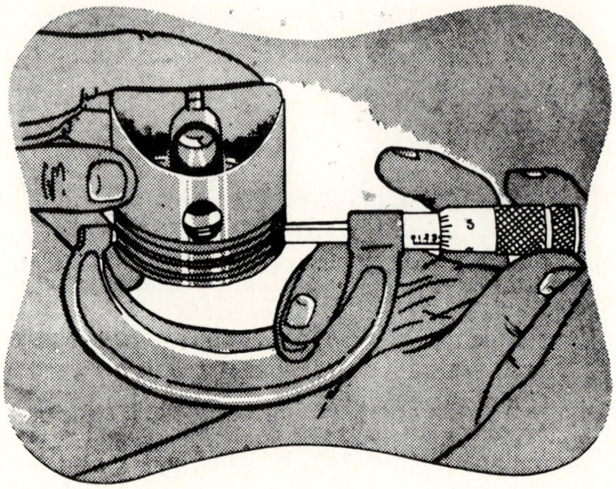

FIGURE 12-9. Measuring ring land diameter (Courtesy of Clinton Engines Corporation).

Checking Ring End Gap

Before installing new rings or reinstalling the old rings (if they are not damaged), you should check the ring end gap. This is a measurement of the distance between the ends of the rings once they have been inserted into the cylinder. If the gap is too great, the ring will not exert sufficient outward force to seal the combustion chamber or to provide adequate oil control (depending on which ring is involved). To check the end gap, first insert the ring into the top of the cylinder. Then, using the inverted piston as a tool, push the ring evenly down into the cylinder. Now remove the piston and measure the ring gap with a feeler gauge. If the gap exceeds the manufacturer's specifications, the ring should not be used. (See Figures 12-10, 12-11, 12-12, and 12-13.) To correct the problem, install an oversize ring. They usually are available in sizes 010 in., 020 in., and 030 in. greater than the original ring size. (*Note*: Pistons are also available in similar oversize ranges. They are used when a badly worn cylinder has been bored or honed out to the next larger size.)

Installing Piston Rings

After the correct piston and piston rings have been selected, the rings must be installed in the ring grooves. The requirements vary from one type of engine to another, so consult the manufacturer's recommendations for particular engines. (See Figures 12-14, 12-15, and 12-16.)

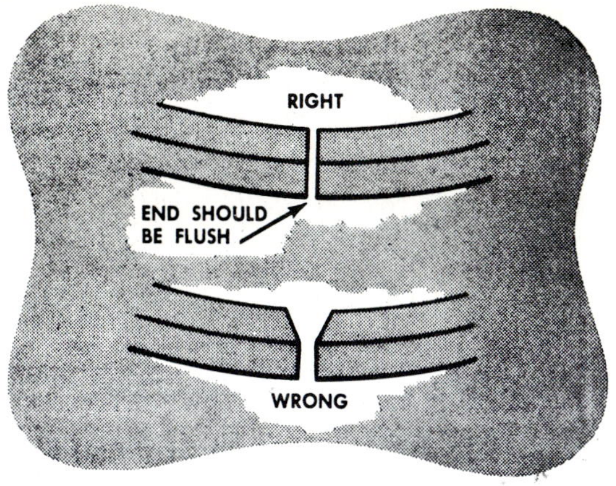

FIGURE 12-10. Good and bad ring gaps (Courtesy of Clinton Engines Corporation).

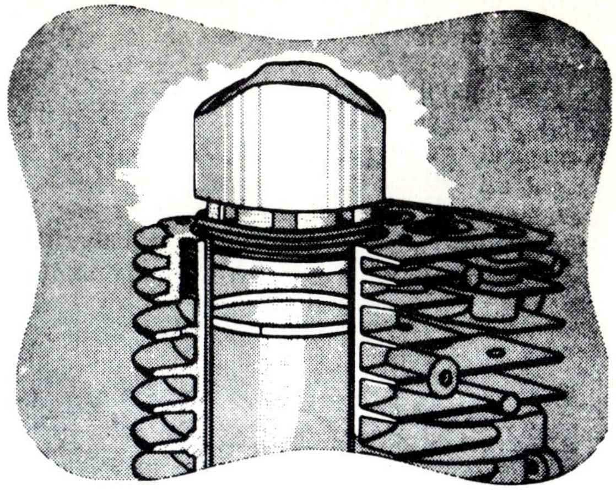

FIGURE 12-11. Positioning a ring in a cylinder for measuring ring gap. (Courtesy of Clinton Engines Corporation).

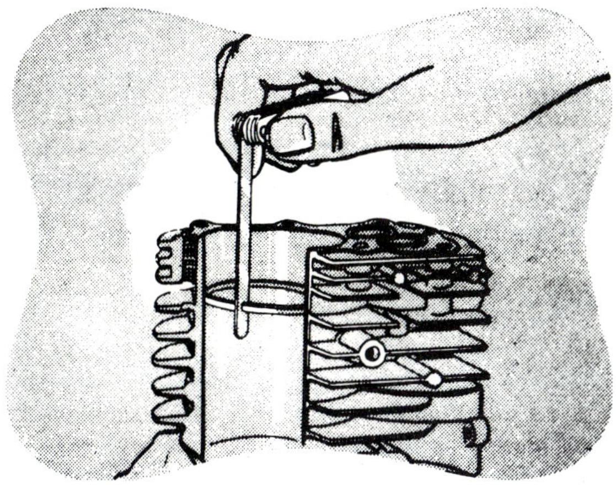

FIGURE 12-12. Measuring ring gap with a feeler gauge (Courtesy of Clinton Engines Corporation).

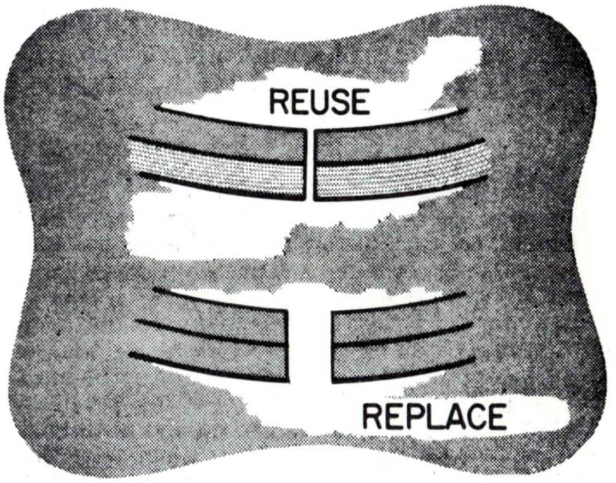

FIGURE 12-13. Ring gap edges should be parallel (Courtesy of Clinton Engines Corporation).

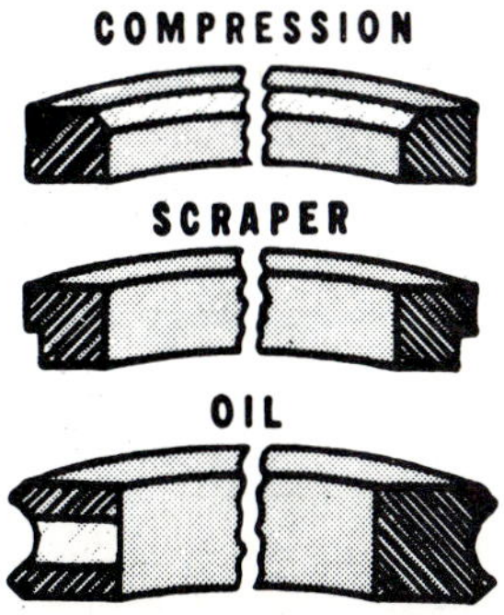

FIGURE 12-14. Typical small engine ring configuration (Courtesy of Clinton Engines Corporation).

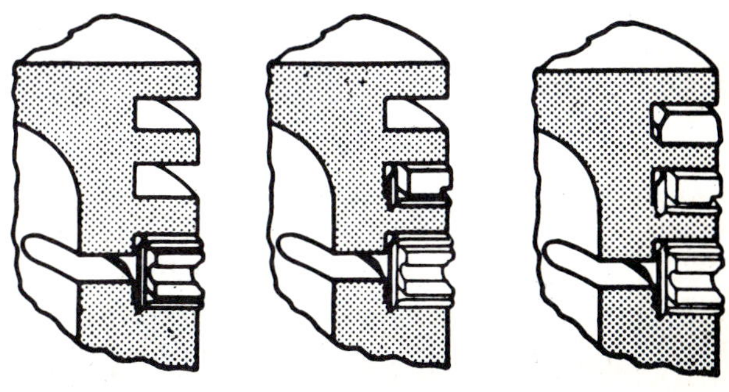

FIGURE 12-15. A view of installed rings (Courtesy of Clinton Engines Corporation).

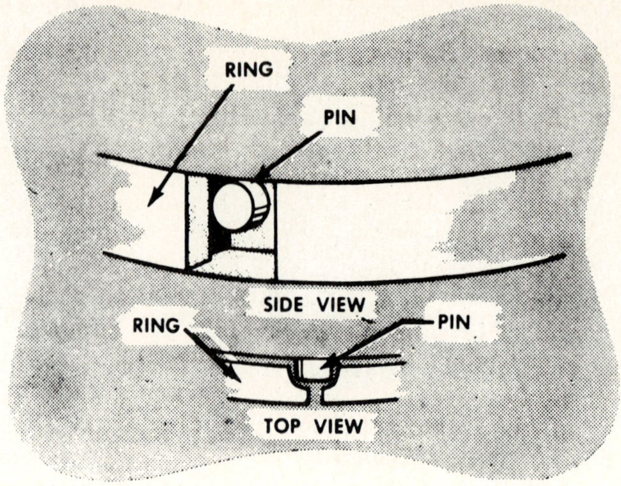

FIGURE 12-16. Pins in the ring groove of some two-cycle engine pistons (Courtesy of Clinton Engines Corporation).

Inspecting the Connecting Rods

An old connecting rod is inspected for one reason only, to see if it can be reused. First, the rod is examined for visibly obvious defects. Some signs to look for are: (1) heat discoloration around the bearing surfaces, (2) scuffing around the bearing surfaces, (3) cracks, and (4) stripped threads—as evidenced by pieces of metal clinging to the cap screws. Any of these conditions are sufficient reason to discard the rod. (See Figure 12-17.)

If the visual inspection did not reveal any defects, then the clearance between the rod's big end and the crankshaft and the clearance between the small end and the wrist pin must be checked.

The rod-to-crankshaft clearance can be determined in several ways. A micrometer and inside hole gauge can be employed, or you can use plastigauge. Here is a general description of the plastigauge technique.

Plastigauge is a small string of plastic material of a known diameter. It comes in an envelope on which is printed a scale for measuring the string. To determine the connecting rod clearances, tear off a small strip of the plastigauge string and place it lengthwise (parallel) on the crankshaft throw. Then replace the connecting rod cap and tighten the screws as if the rod were going to be reinstalled. As the cap is tightened, the plastigauge string will be permanently deformed. After the cap has been tightened to the specified torque level, it is removed and the plastigauge string is retrieved. The width of the flattened string is then compared to the scale on the envelope. This reading gives the clearance between the rod and shaft. (See Figures 12-18, 12-19, and 12-20.) If the reading exceeds the recommended clearance, the connecting rod should be discarded in most cases.

Note: The rod to wrist pin clearance must obviously be checked with a micrometer and inside hole gauge. If this reading is out of specification, it may be possible, depending on the engine, to use an oversize wrist pin.

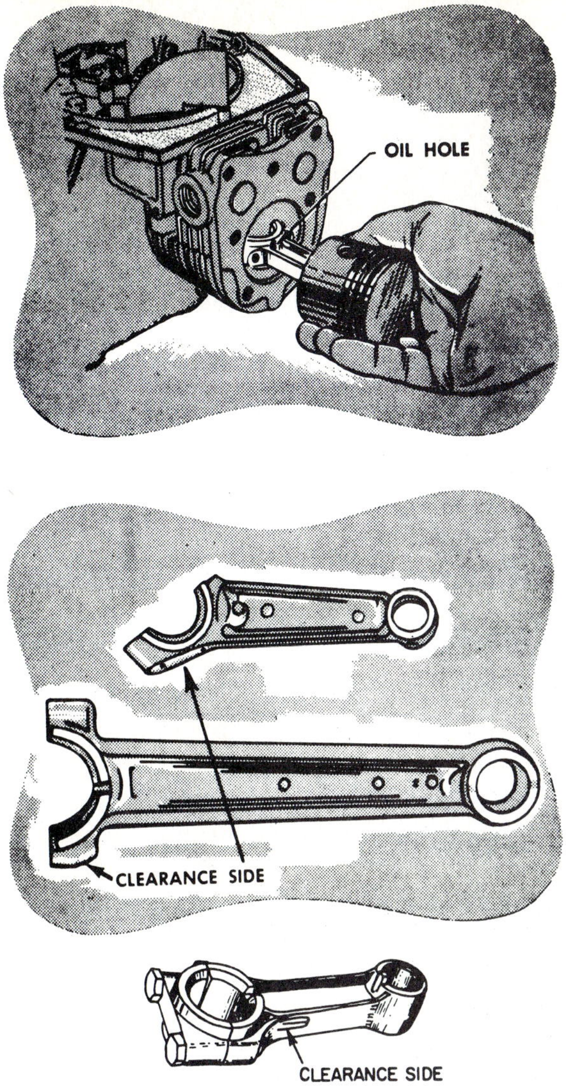

FIGURE 12-17. Inspection of connecting rods for oil hole position, cracks, and imperfections (Courtesy of Clinton Engines Corporation).

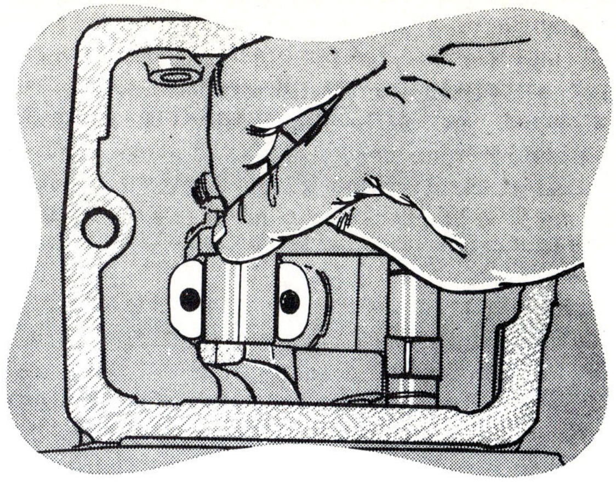

FIGURE 12-18. Placing plastigauge to check for rod-to-shaft clearance (Courtesy of Clinton Engines Corporation).

FIGURE 12-19. Placing plastigauge to check shaft for taper (Courtesy of Clinton Engines Corporation).

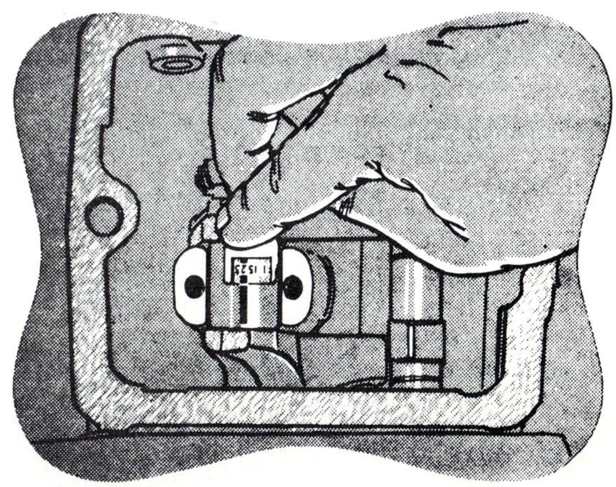

FIGURE 12-20. Checking the width of a plastigauge, using the scale from the package (Courtesy of Clinton Engines Corporation).

Inserting the Piston into the Cylinder

After the connecting rod has been checked and reattached to the piston, the piston is inserted into the cylinder. Several points to bear in mind during this operation are

1. *Direction of installation.* Although some pistons may be installed from either direction into the cylinder (so long as the piston head is up), others must be installed in a particular direction. Check the manufacturer's specifications for particular engines.
2. *Position of connecting rod.* As we noted in the chapter on lubrication, most connecting rods have an oil access hole. It is of vital importance to make sure that this hole is pointing in the correct direction.
3. *Compression of the rings.* Since the expanded rings will project out further than the cylinder walls, they must be compressed before the piston can be installed. Several types of ring compressors are available for this job. Also, to protect the bore and rings during the first few moments of engine operation, a light coat of engine oil should be applied to the cylinder walls. (See Figures 12-21 and 12-22.)

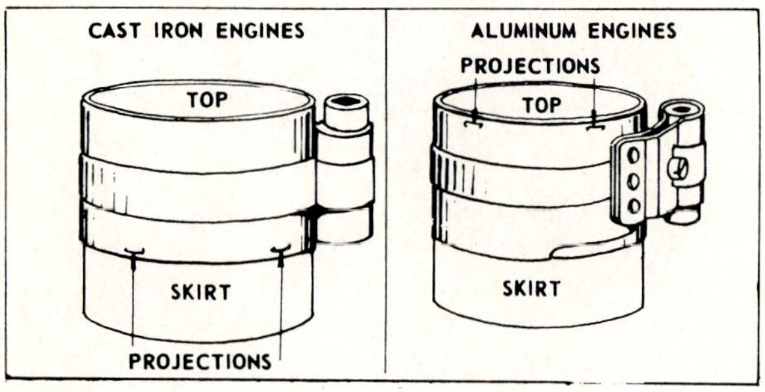

FIGURE 12-21. Typical piston ring compressor (Courtesy of Briggs & Stratton Corporation).

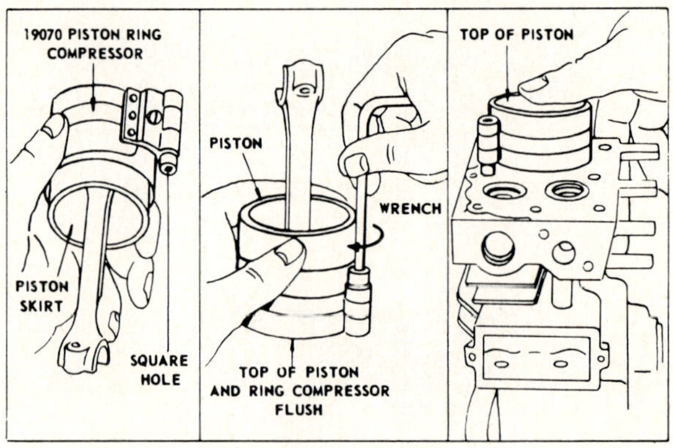

FIGURE 12-22. Installing the ring compressor and inserting the piston into the cylinder (Courtesy of Briggs & Stratton Corporation).

Installing the Connecting Rod Cap

Follow these general steps when installing the connecting rod caps:

1. Before doing anything, make sure that the crankshaft journal is clean. Wipe it with a soft, clean cloth, then apply a light coating of clean lubricating oil.
2. The connecting rod caps are machined to fit the rod ends in only one way. Make sure that the rod caps are installed in the correct direction.
3. Install the oil dippers (if applicable), then torque the cap screws to the recommended levels. Lock the screws into position. (See Figures 12-23 and 12-24.)
4. Finally, turn the crankshaft several times by hand to make sure that there is no binding or excessive drag.

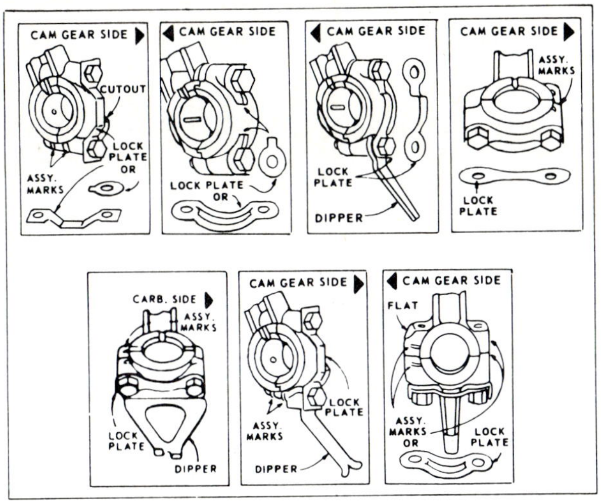

FIGURE 12-23. Oil dippers and rod cap lock plates (Courtesy of Briggs & Stratton Corporation).

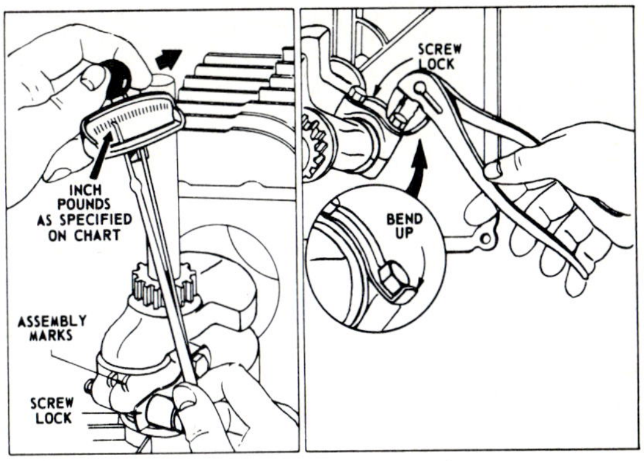

FIGURE 12-24. Torquing the connecting rod cap screws and bending the screw locks (Courtesy of Briggs & Stratton Corporation).

13

ENGINE BLOCK, CYLINDER, AND CRANKSHAFT SERVICE

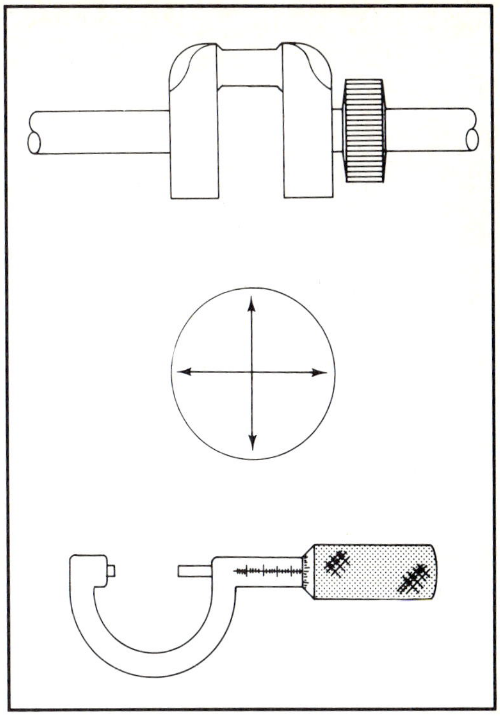

GENERAL

Work performed on the engine block, cylinder, and crankshaft is considered major service. Such work is time consuming and expensive and requires considerable skill. For many small engines, major service will never be done because the cost of the repairs exceeds the value of the engine. When it is performed, major service is usually carried out by trained technicians in a major repair shop. For these reasons, we will concentrate on basic repair principles, rather than detailed procedure for specific engines. (See Figure 13-1.)

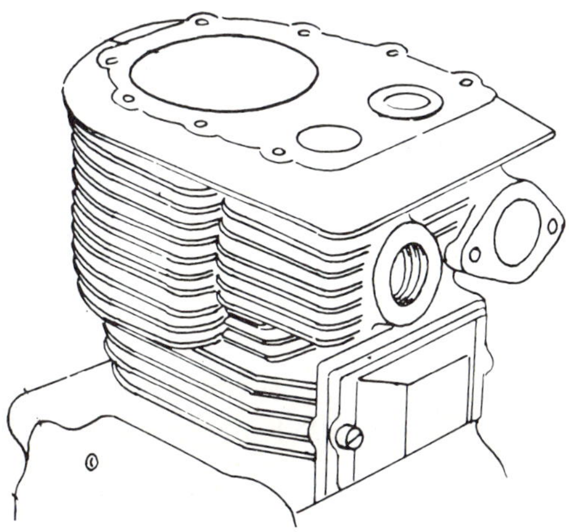

FIGURE 13-1. An engine block (Courtesy of Briggs & Stratton Corporation).

TOOLS

Performing major service requires skill with some specialized tools and equipment. For instance, you must know how to use micrometers and dial indicators to measure cylinder wear. You must also be able to use various types of boring and honing machines to resize the cylinders when your measurements reveal excessive wear. Chapter 16 on measuring tools introduces you to micrometers. This chapter will cover some general aspects of boring and honing machines. If you are in a school program, these devices will probably be covered in more detail. If you are on your own, perhaps thinking about setting up your own operation, it would be a good idea to spend some time in a shop getting familiar with this equipment. (See Figures 13-2, 13-3 and 13-4.)

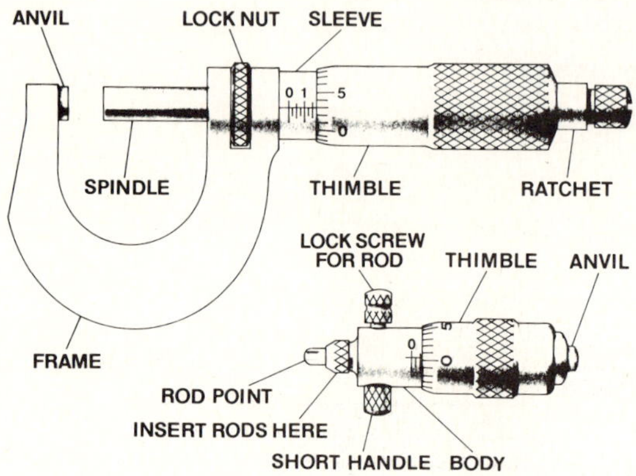

FIGURE 13-2. Inside and outside micrometers (Courtesy of Gould, Inc.)

FIGURE 13-3. Small-engine cylinder boring bar (Courtesy of Clinton Engines Corporation).

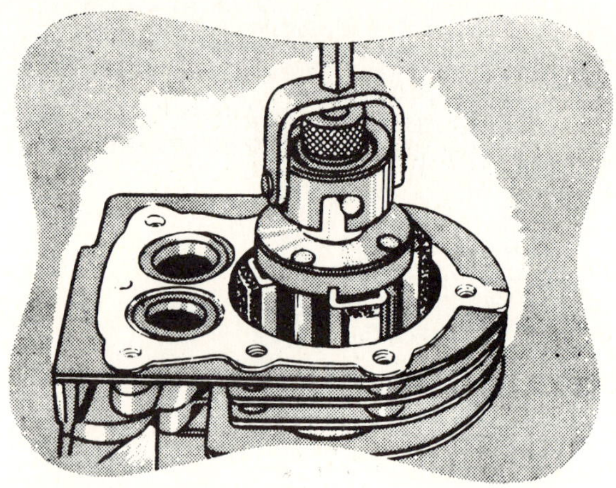

FIGURE 13-4. A cylinder hone (Courtesy of Clinton Engines Corporation).

DISASSEMBLING THE ENGINE

As we have indicated before, common sense is the rule when taking an engine apart. Don't hurry; follow the manufacturer's recommendations and use the correct tools.

CLEANING THE COMPONENTS

Common sense should also be your guide when cleaning the components. Use a stiff wire brush and solvent on metal parts and a milder detergent on plastic and rubber components. After you clean everything, allow the pieces to dry. (Use an air gun if you have one.)

VISUAL INSPECTION

It is very important to examine all the disassembled engine components. Some conditions to look for include:

1. Cracks, gouges and discolorations
2. Metal filings in the oil sump
3. Damaged screws and threaded holes

After you have completed your visual inspection, put the parts that need replacement in one pile. In that way, none will be overlooked when you make your trip to the parts store.

Cylinder Measurement

Even though the visual inspection may not reveal any problems with the cylinder, it must still be checked with measuring instruments. Your eye will not be able to tell if the cylinder has been worn too large, or if it has become tapered or out of round.

Several instruments can be used to measure the cylinder diameter. The cylinder gauge, for instance, is a dial micrometer specially adapted for measuring cylinder diameters. A general-purpose, inside micrometer may also be employed. However, a telescoping gauge used in conjunction with an outside micrometer is the most common method for determining cylinder wear. (See Figure 13-5.)

To use a telescoping gauge, first place the spring loaded, "T"-shaped end inside the cylinder and then loosen the thumb screw on the handle.

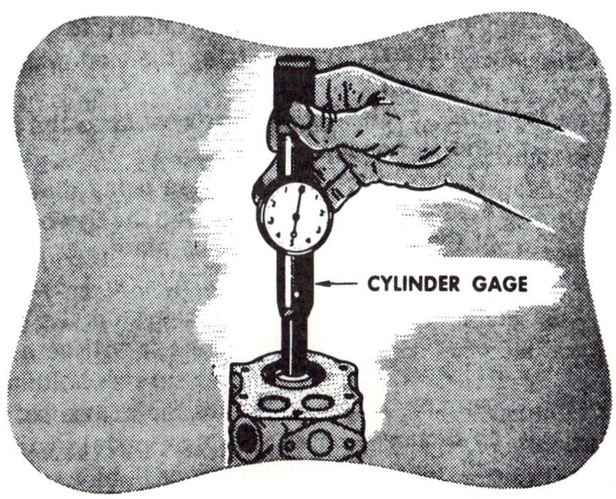

FIGURE 13-5. A cylinder dial gauge (Courtesy of Clinton Engines Corporation).

The ends of the "T" will push out toward the sides of the cylinder. Once the cylinder diameter is established, tighten the thumb screw and remove the gauge. Now measure the extended ends with a micrometer. To be safe, the cylinder should be measured at the six locations indicated in Figure 13-6.

Three conditions are checked for wear:

1. *Excessive diameter.* If the diameter is worn more than 0.003 in. at any measuring point, the cylinder will need resizing.
2. *Taper.* This refers to diameter differences from the top to the bottom of the cylinder. The minimum allowable taper varies from engine to engine, so you must check the manufacturer's specifications for particular engines.
3. *Out-of-round.* This refers to diameter differences encountered when you make various measurements at the same height or level in the cylinder. Most manufacturers recommend resizing the cylinder if the differences exceed 0.003 to 0.005 in. (See Figure 13-7.)

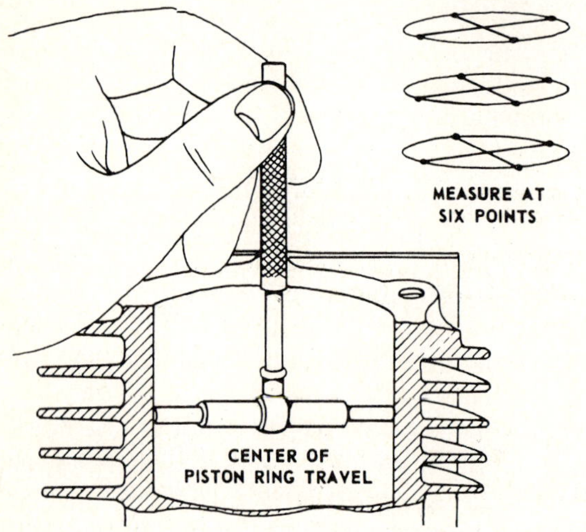

FIGURE 13-6. Measuring a cylinder with a telescoping hole gauge (Courtesy of Briggs & Stratton Corporation).

FIGURE 13-7. A telescoping hole gauge (Courtesy of Clinton Engines Corporation).

Determining the Cylinder "Oversize"

If the cylinder measurements are out of specification, or if the visual inspection reveals a gouge in the cylinder wall, then the cylinder must be resized. The term "resizing" is used instead of "reboring" to emphasize the fact that the cylinder is not just drilled out. Since replacement pistons are available only in certain sizes, the cylinder diameter must be increased by a specified amount. Generally, three levels of oversize pistons are available–0.010 in., 0.020 in., and 0.030 in. greater than the standard size. Usually, the oversize level is indicated on the head of the piston by a 10, 20 or 30 notation. The particular oversize for a given cylinder depends on the degree of wear–the greater the wear, the larger the oversize required. After the oversize is determined, the amount is added to the stock bore. For instance, if a 2.375 in. standard cylinder requires a 0.010 oversize, to correct wear, you would increase the existing diameter by 0.010 in. The new diameter of 2.385 in.

would then allow a 0.010 oversize piston to operate with the proper clearance. (See Figure 13-8.)

Several methods can be used to resize cylinders. The two that we shall discuss here are the cylinder hone and the boring bar methods.

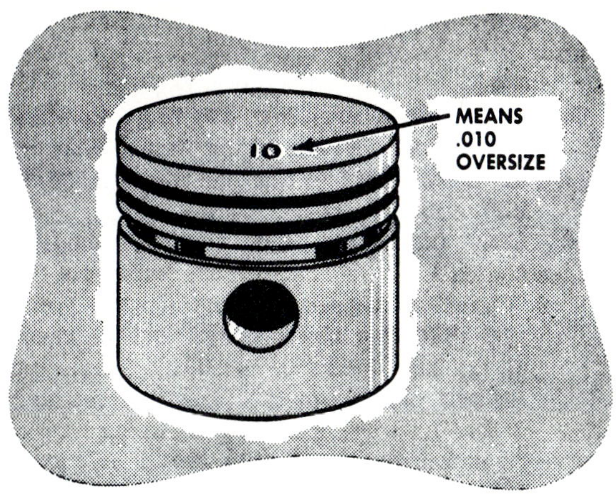

FIGURE 13-8. Oversize piston identification (Courtesy of Clinton Engines Corporation).

Resizing with a Cylinder Hone

A cylinder hone is a drum-shaped grinding wheel mounted on a drill press or portable drill. Abrasive stones attached to the wheel gradually cut away metal from the cylinder to obtain the required oversize. Coarse stones are used at the beginning of the operation to remove larger amounts of metal. Toward the end of the procedure, finer stones are attached to the hone. Fine stones are also used to remove minor scratches from or to deglaze an otherwise satisfactory cylinder. The following steps outline some general procedures for using a cylinder hone. (See Figure 13-9.)

1. Carefully clean the table of a conventional drill press and mount the cylinder block on the table. Make sure that the block is steady and does not rock back and forth. (*Note*: If a drill press is not available, a one-half-inch, hand-operated drill will do.)
2. Mount the cylinder hone in the drill press chuck and set the spindle speed to 450–600 RPM. (See Figure 13-10.)
3. Center the cylinder block under the drill spindle.
4. Install a set of coarse stones in the hone.
5. If the drill press has adjustable stops, set them so that the hone protrudes ½ to 1 in. from the top and bottom of the cylinder at both ends of its travel. If you must regulate the travel by hand, try to stay within these bounds.
6. Move the hone to the narrowest part of the cylinder, and then turn the adjusting nut until the stones touch the cylinder wall. Although judging hone tension is a matter of experience, you must be able to turn the hone by hand before beginning the operation. If you can't, the hone is too tight.
7. Turn the drill on and start moving the hone up and down from one end of

its travel to the other. Try to maintain a rate of about 40 cycles per minute to obtain the required crosshatch pattern pictured in Figure 13-11.

8. When you start the honing operation, the drill motor will likely slow down as the hone passes through the narrow part of the cylinder. Then, as the size nears the desired diameter, the drill speed will remain uniform up and down its travel. As soon as this happens, stop the drill and measure the cylinder. When the size is within 0.015 in. to 0.002 in. of the desired diameter, install fine stones on the hone, and finish the operation.

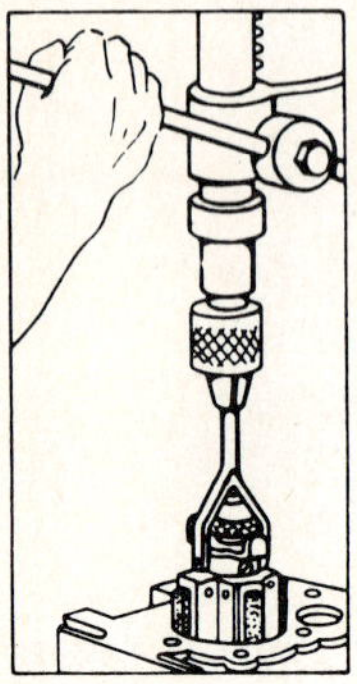

FIGURE 13-9. Using the cylinder hone (Courtesy of Briggs & Stratton Corporation).

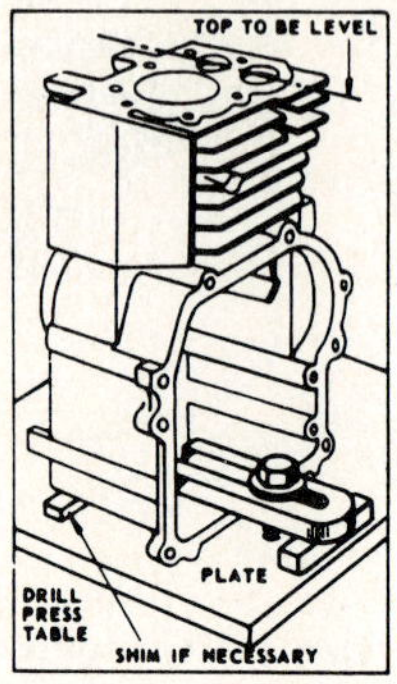

FIGURE 13-10. Mounting the small engine block in a drill press (Courtesy of Briggs & Stratton Corporation).

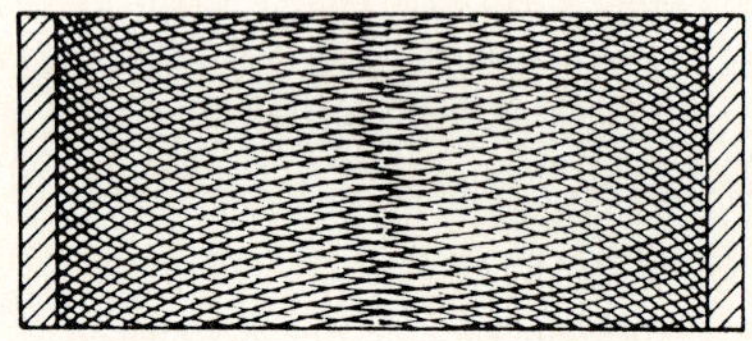

FIGURE 13-11. Crosshatch cylinder surface finish (Courtesy of Briggs & Stratton Corporation).

Resizing the Cylinder with a Boring Bar

Some boring bar machines resemble large, complex drill presses. However, instead of a conventional drill bit, the boring bar uses a bar-shaped cutting element with replaceable knife edges. Most boring bars provide a number of special features to make the cylinder sizing operation easier and faster. A built-in micrometer is used to determine the depth of the cut. Centering devices and automatic feeds are also common. The finish provided by some boring bars requires no further service. However, most experienced technicians recommend that you obtain the final finish on the cylinder with a fine stone hone. (*Note*: Some authorities also recommend cutting a 45-deg bevel on the top of the cylinder so that the rings and piston can be more easily installed. (See Figure 13-12.)

Cleaning the Engine after Resizing

No matter what technique is employed, the cylinder should be cleaned after a resizing operation. Mop the cylinder walls thoroughly with soap and water to make sure that no loose particles remain. (See Figure 13-13.)

Crankshaft Service

Always examine the crankshaft before putting it back into the engine. (See Figures 13-14 and 13-15.) Some conditions to look for include:

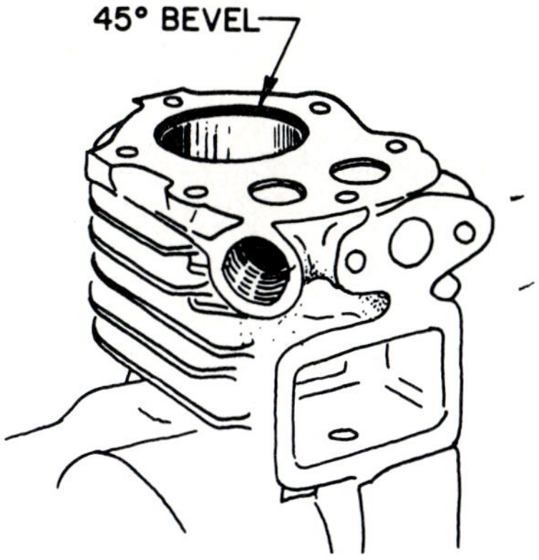

FIGURE 13-12. Forty-five-degree bevel at the top of a cylinder (Courtesy of Clinton Engines Corporation).

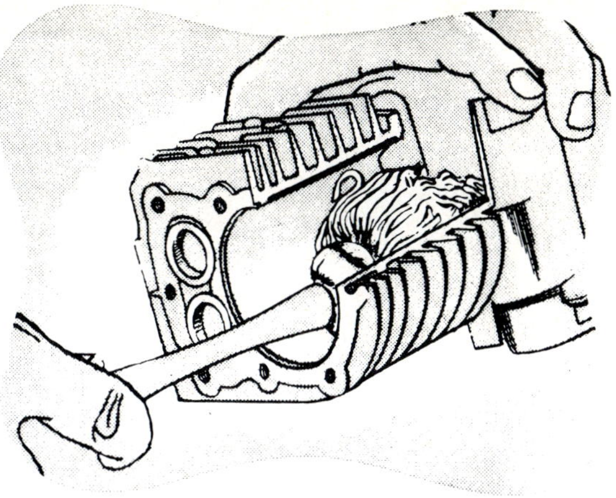

FIGURE 13-13. Cleaning out a resized cylinder (Courtesy of Clinton Engines Corporation).

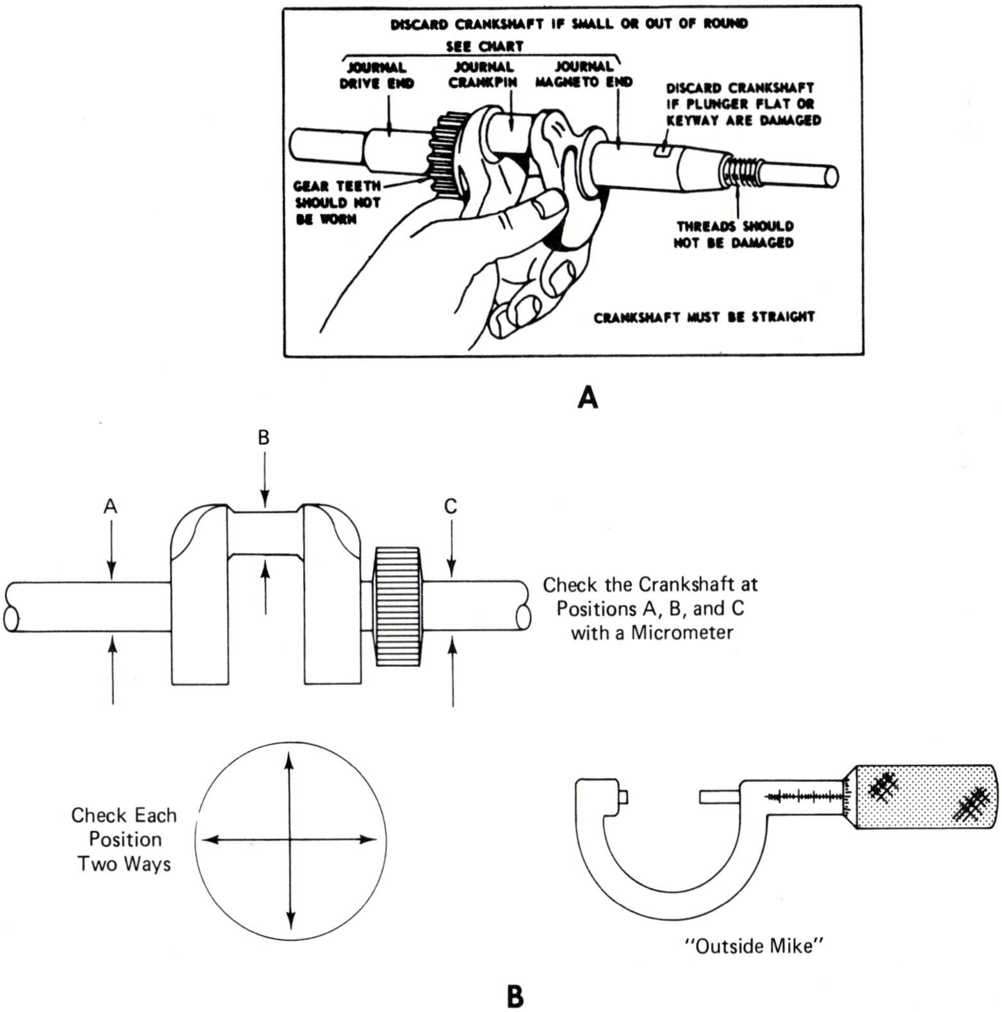

FIGURE 13-14. (A) Crankshaft checks; (B) using a micrometer to check the crankshaft (Courtesy of Briggs & Stratton Corporation).

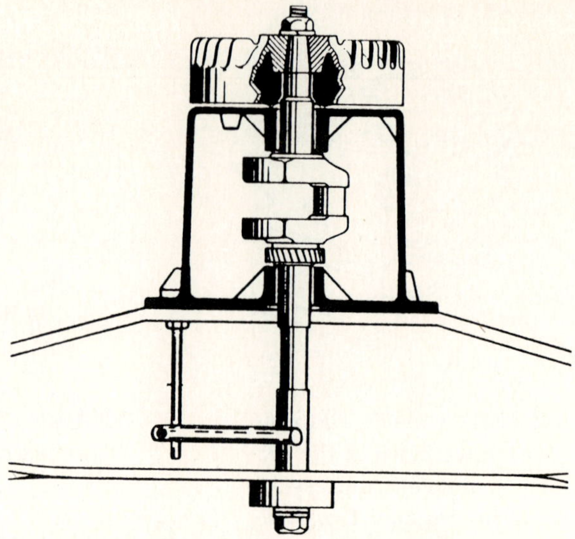

FIGURE 13-15. Checking for a bent crankshaft (Courtesy of Clinton Engines Corporation).

1. Signs of gross physical damage, such as cracks, misalignments, etc.
2. Discoloration (usually blue in color) indicating heat damage.
3. Rough main and rod bearing surfaces.
4. Damaged keyways and threads.

Since small engine crankshafts are usually more expensive to repair than replace, any of these conditions will usually justify discarding the crankshaft.

After the crankshaft has been visually inspected, it should be measured with a micrometer. Again, if the results are out of specifications, the crankshaft should probably be replaced.

Although crankshafts on small, inexpensive engines are usually replaced rather than repaired, there is one common failure that can be serviced. When the blade of a rotary lawnmower encounters an immovable object (rock stump, etc.), the inertia built up in the flywheel has a tendency to twist the crankshaft into a pretzel shape. Fortunately, before this can happen, the soft key that connects the flywheel to the crankshaft will break. All you have to do to remedy this problem is reinstall the flywheel, using a new key.

Installing the Crankshaft

A primary consideration when installing the crankshaft is to make sure the oil seals at both ends of the shaft fit properly. Use a seal driving rod that is long enough to fit over the end of the crankshaft and still engage the seal. Also make sure that the heavy lip of the oil seal faces the crankcase. Place sealant only on the outer circumference of the seal. Remember to be careful when removing the seal in the first place. It is easy to damage the main bearing hole. (See Figure 13-16.)

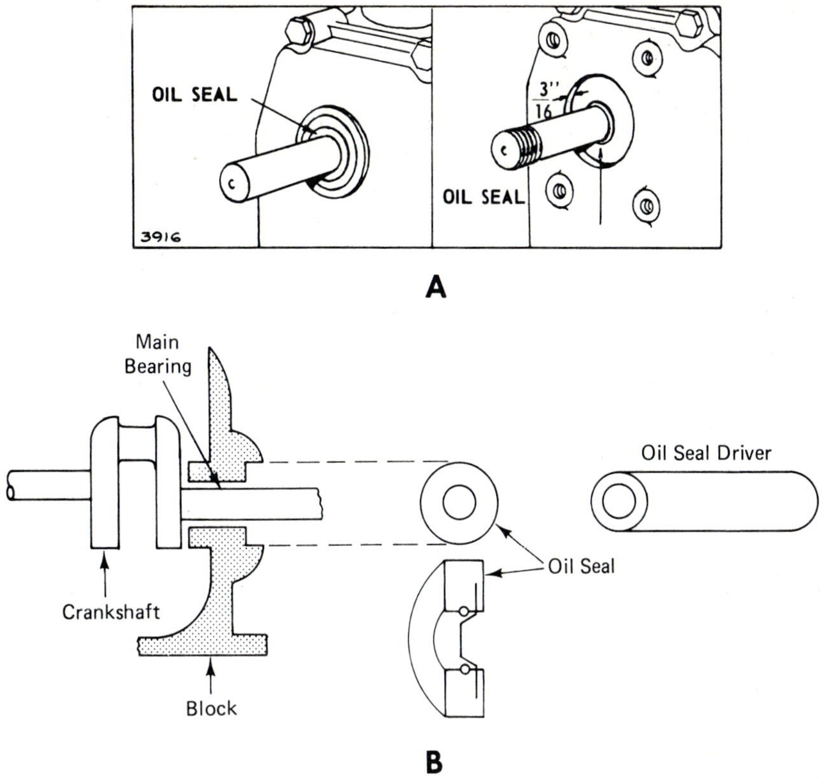

FIGURE 13-16. (A) Oil seal location; (B) replacing oil seals (Courtesy of Briggs & Stratton Corporation).

Small Engine Main Bearings

The main bearings are located in the holes through which the ends of the crankshaft project. Most inexpensive small engines use the actual holes as the main bearings. Badly worn main bearing holes are drilled and line reamed, then fitted with bushings. The bushings are drilled to accommodate the crankshaft. When this type of service is not possible, the block or sump plate must be replaced. (See Figure 13-17.)

Some "larger" small engines use roller-type main bearings. They are adjusted by adding or removing shim stock from the main bearing supports.

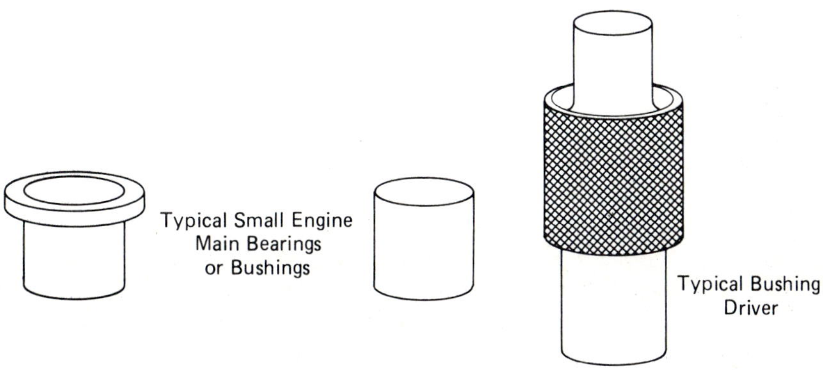

FIGURE 13-17. Main bearing bushings.

Usually adjustment is not needed until wear takes place. Some types of ball bearings are press fit on the crankshaft and can be removed only with a special puller or press. (See Figures 13-18, 13-19, and 13-20.)

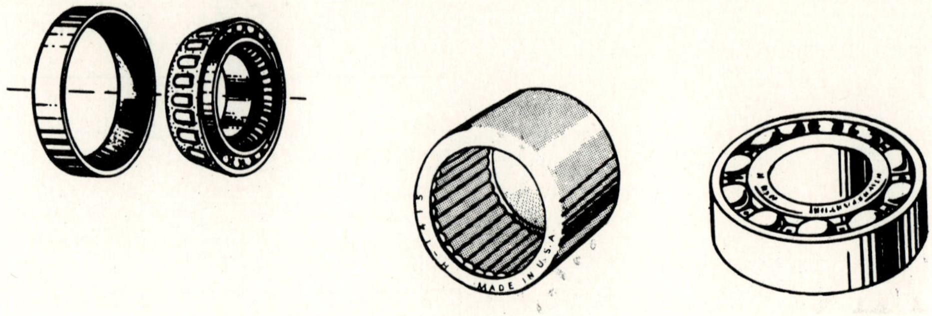

FIGURE 13-18. Small-engine bearings (Courtesy of Clinton Engines Corporation).

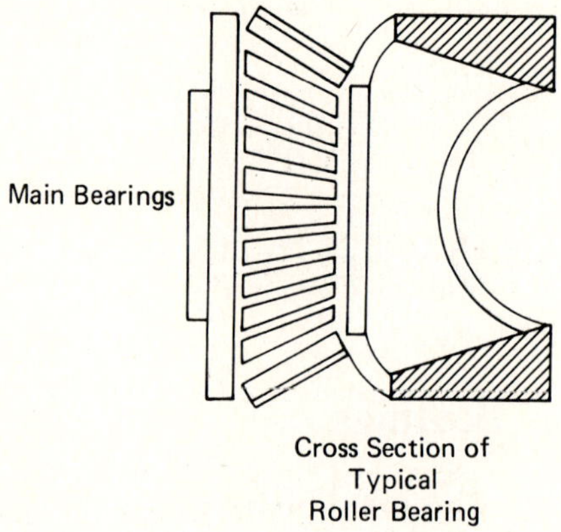

FIGURE 13-19. Cross-section of a bearing.

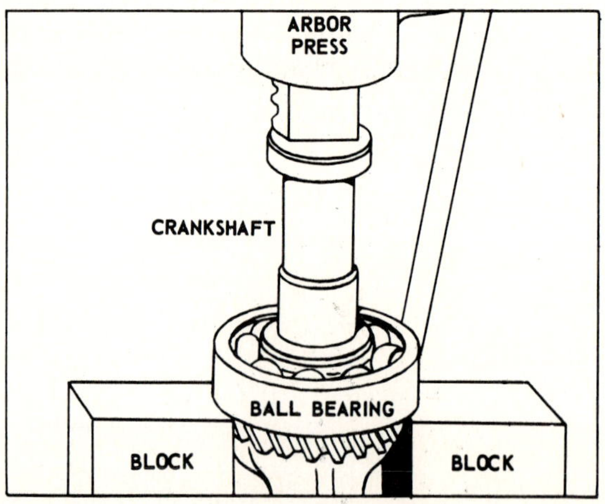

FIGURE 13-20. Installing bearings with an arbor press (Courtesy of Briggs & Stratton Corporation).

14

ROUTINE MAINTENANCE AND TUNE-UP

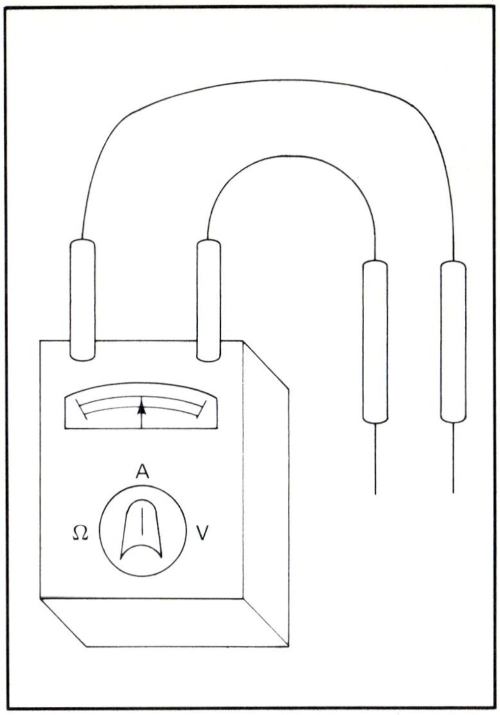

GENERAL

This chapter examines routine maintenance and tune-up procedures typically performed on small gasoline engines. Whenever possible, detailed directions are given for specific tasks. However, considering the variety and number of small engines, it is always a good idea to consult the manufacturer's reference materials before attempting to do work on your own.

This chapter presumes that you have already studied the rest of the book and are familiar with the basic operation of small engines. If you have a question, refer to one of the other chapters.

ROUTINE MAINTENANCE

Routine maintenance varies from manufacturer to manufacturer and with the type of service for which an engine is being used. In most cases, though, the following components need to be routinely examined to make sure that the engine remains in satisfactory condition.

Fuel Storage and Delivery

FUEL STRAINER

Most fuel tanks have a wire mesh strainer located at the outlet of the fuel tank. (See Figure 14-1.) Remove and clean the strainer on a periodic basis. Don't wait until the fuel line clogs up and the engine stops running.

FUEL FILTER

Most fuel filters are disposable paper cartridges located in the line between the tank and carburetor. This type of fuel filter should be replaced at regular intervals. Some fuel systems also include a glass filter bowl mounted beneath the gas tank. (See Figure 14-2.) You can tell by looking at the sediment chamber at the bottom of the bowl how much water and trash have accumulated. The bowl should be removed and cleaned as necessary.

FUEL LINES

Rubber or plastic hoses should be replaced when they become hard or brittle. The end fittings on all delivery hoses and pipes should be checked periodically and replaced or tightened as required. Leaks not only waste fuel, they can cause fires.

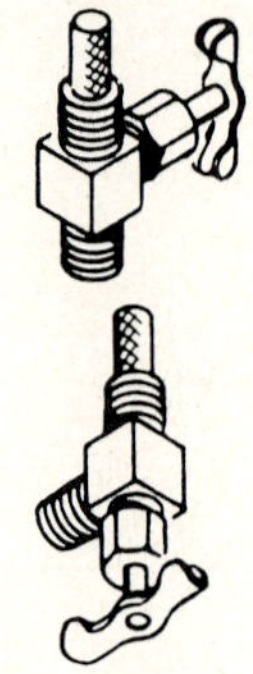

FIGURE 14-1. Fuel tank strainers (Courtesy of Clinton Engines Corporation).

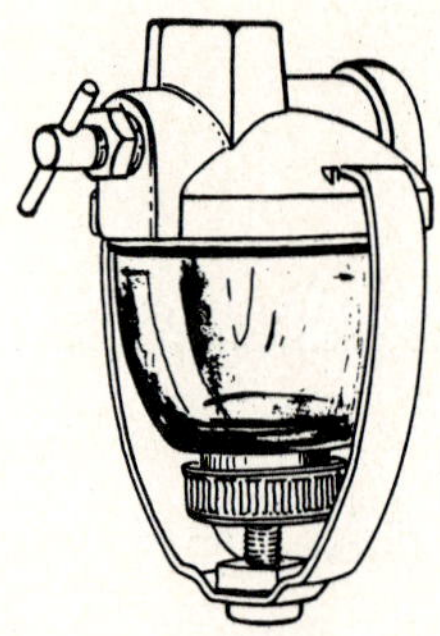

FIGURE 14-2. Sediment-bowl fuel filter (Courtesy of Clinton Engines Corporation).

FUEL TANK

If subjected to heavy vibrations or contaminated fuel, the fuel tank should be removed and inspected. Otherwise, little or no regular maintenance is required. (See Figure 14-3.)

FUEL TANK CAP

Make sure that the vent hole in the gas cap isn't blocked. Fuel won't flow from the tank if the vent is blocked. Also, check the gas cap gasket from time to time. (See Figure 14-4.)

Lubricating System

It is very important to follow the manufacturer's recommendations regarding lubrication system maintenance. For four-stroke engines, always make sure that the crankcase contains the correct amount of the proper fuel. Also, be sure to drain and replace the crankcase oil on schedule. If the engine is operated in a particularly dirty environment, change the oil more often. (See Figures 14-5, 14-6, 14-7, 14-8, and 14-9.)

The primary lubrication considerations for two-stroke engines are to:

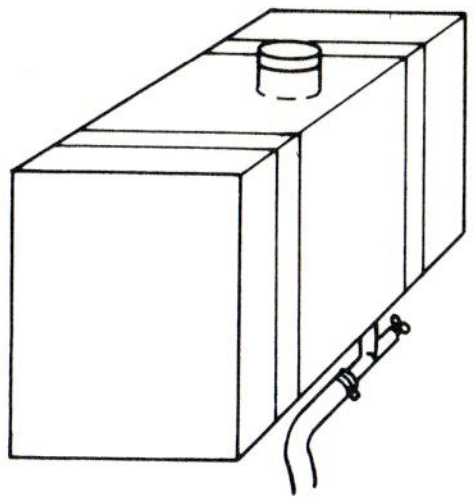

FIGURE 14-3. Small-engine fuel tank.

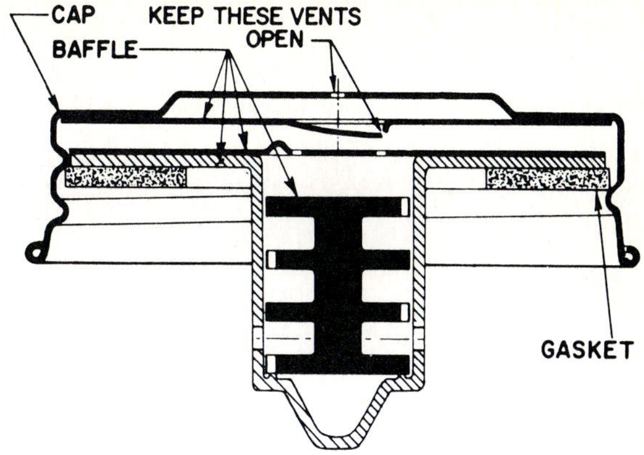

FIGURE 14-4. Fuel tank cap (Courtesy of Clinton Engines Corporation).

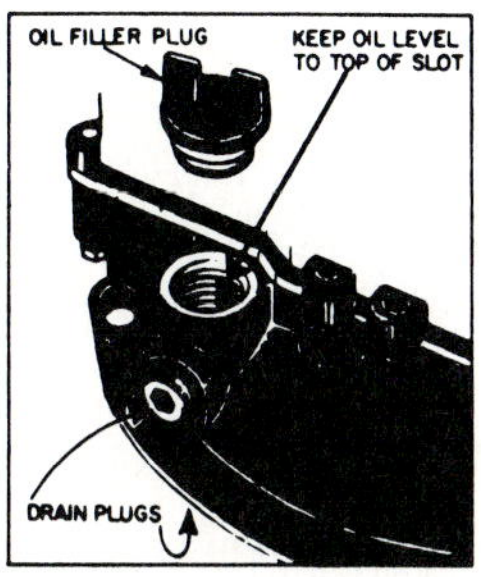

FIGURE 14-5. Oil filler, drain, and level plugs (Courtesy of Clinton Engines Corporation).

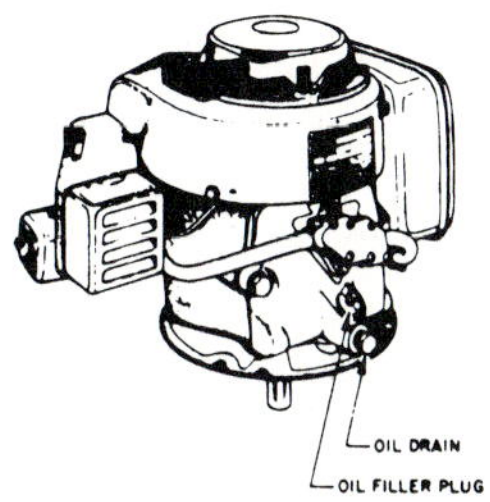

FIGURE 14-6. Vertical-shaft, small-engine oil filler plug (Courtesy of Clinton Engines Corporation).

FIGURE 14-7. Horizontal-shaft, small-engine dip stick (Courtesy of Clinton Engines Corporation).

FIGURE 14-8. Horizontal-shaft, small-engine oil filler plug (Courtesy of Clinton Engines Corporation).

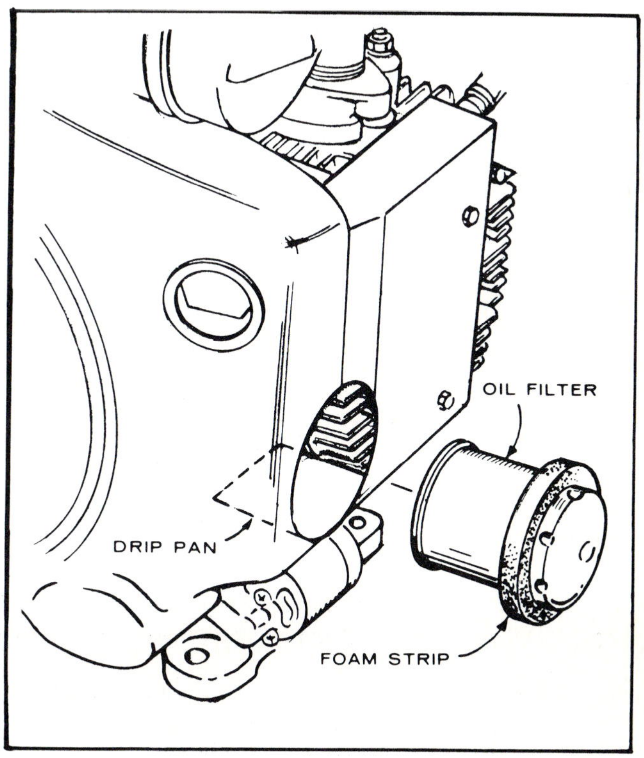

FIGURE 14-9. Oil filter (Courtesy of Onan).

(1) mix the oil and gasoline in the correct amounts, and (2) ensure that only the two-stroke oil recommended by the manufacturer is used in the engine.

Note: Some "larger" small engines (two cylinders and up) may contain an oil filter. If so, it should be changed on schedule.

Air Cleaner

An improperly fitted air cleaner can allow dirt to pass and thereby shorten the life of the engine. A clogged or dirty air cleaner will restrict the air flow and cause the fuel mixture to become rich enough to foul the spark plugs. Therefore, it is important to maintain a clean, properly fitted air cleaner.

Generally, three types of air cleaners may be found on small engines:

FOAM ELEMENT

May be reused after being cleaned in solvent and dried. Some manufacturers suggest squirting a light coat of oil on the foam to help trap dirt. (See Figures 14-10 and 14-11.)

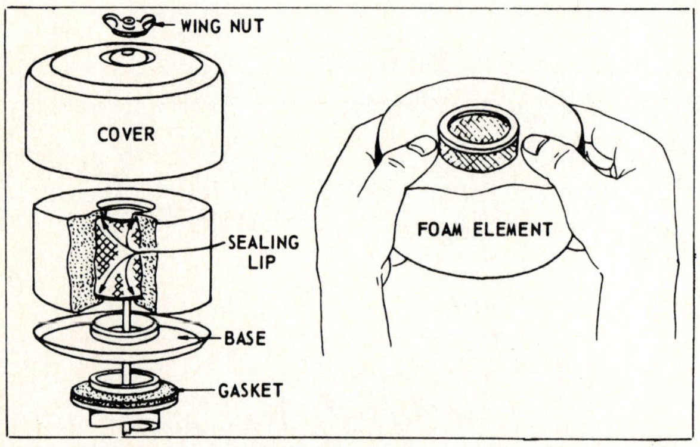

FIGURE 14-10. Oil foam air cleaners (Courtesy of Briggs & Stratton Corporation).

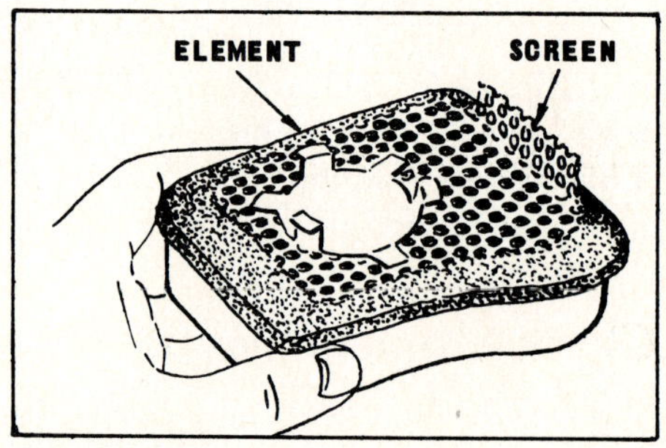

FIGURE 14-11. Foam air cleaner element (Courtesy of Briggs & Stratton Corporation).

PAPER ELEMENT

A disposable, pleated insert. Some technicians recommend removing paper filters between regular change intervals and blowing off any dirt that may have accumulated. (See Figure 14-12.)

OIL BATH

Used primarily on older-style engines. Incoming air bubbles through an oil bath on its way to the carburetor (leaving dirt behind in the oil bath). This is a good type of filter if the oil is changed regularly, worthless if it is not. (See Figure 14-13.)

Battery

If an engine has a self-starter, it may also likely have a battery that requires regular maintenance. Keep the battery case and terminals clean. If the bat-

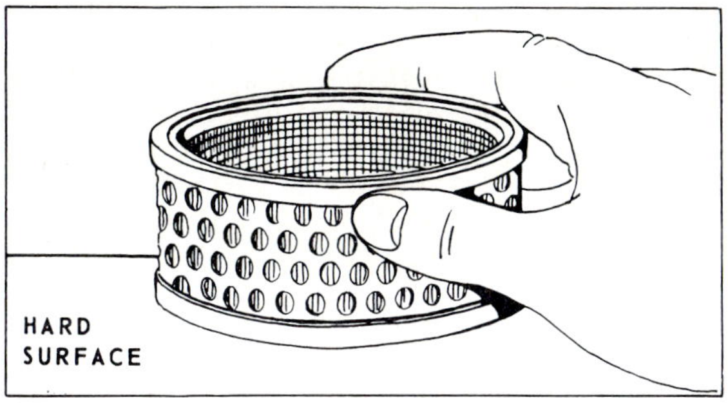

FIGURE 14-12. Paper element air cleaner (Courtesy of Briggs & Stratton Corporation).

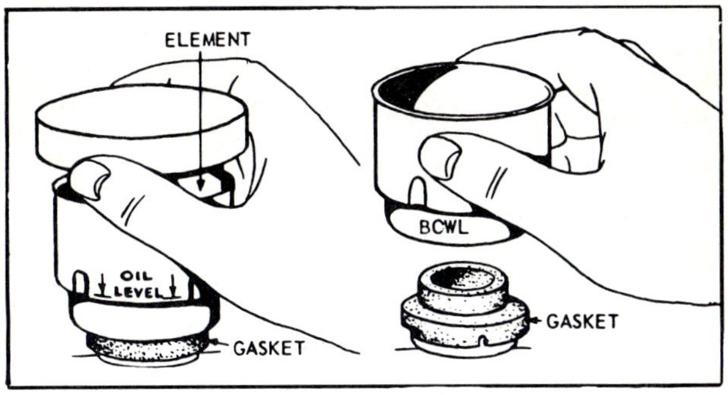

FIGURE 14-13. Oil bath air cleaner (Courtesy of Briggs & Stratton Corporation).

tery is lead acid-type with accessible cells, the water level should be checked from time to time and replenished as needed. (See Figure 14-14.)

TROUBLESHOOTING

Before performing a tune-up, it is a good idea to troubleshoot the engine. It can save unnecessary parts replacement and potentially hours of wasted time.

Following are some basic troubleshooting techniques. The extent of the service performed will depend in large measure on the results of these tests.

Check Magneto Output

Magnetos must supply at least 7000 volts to produce an arc across a 0.028 in. gapped spark plug. To be on the safe side, most authorities suggest checking for an arc on a plug that has been gapped wider than normal. In one example, an 18 mm spark plug is adjusted so that the gap of 0.156 of an

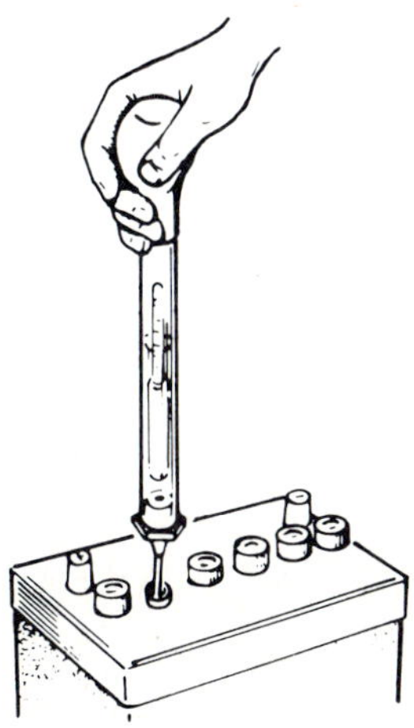

FIGURE 14-14. Checking a battery with a hydrometer (Courtesy of Onan).

inch (or $^5/_{32}$ in.) exists between the center and the ground electrodes. Then the plug is connected to the high tension lead and held against ground while the engine is cranked. If a spark appears, it means that the magneto is producing approximately 10,000 volts. If a spark is not present, the magneto and its components should be examined before any other testing or repairs are attempted. (See Figure 14-15.)

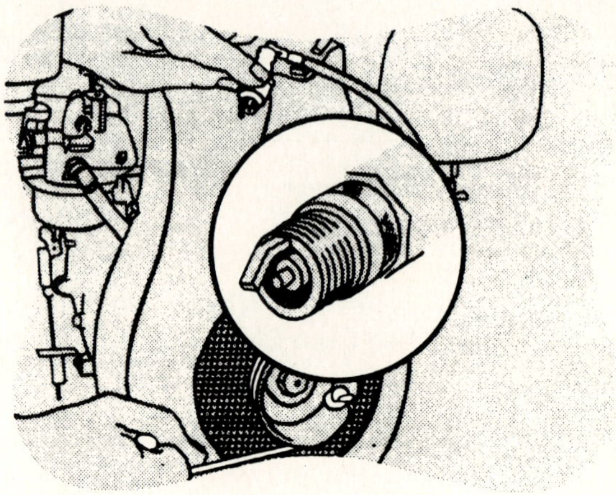

FIGURE 14-15. Wide-gap spark plug test (Courtesy of Clinton Engines Corporation).

Compression Check

The compression or pressure produced in the combustion chamber of a single cylinder is very important. If that one cylinder fails, there are no others to fall back on. Here is a generally accepted procedure for checking compression (see Figure 14-16.):

1. Remove the spark plug from the cylinder.
2. Insert a compression gauge in the spark plug hole. Either a screw-in or hand-held gauge will work.
3. While turning the engine over at normal cranking speed, observe the reading on the gauge.

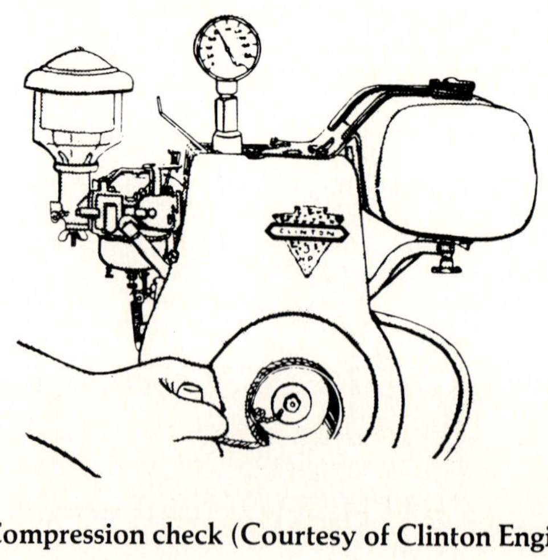

FIGURE 14-16. Compression check (Courtesy of Clinton Engines Corporation).

4. The reading should be above 60 psi (pounds per square inch) for two-stroke engines and above 65–70 psi for four-stroke engines (which produce less than 4½ hp). To be sure, check the manufacturer's specifications for the engine you are working on.
5. If the reading is below specification, you may need to repair or replace one or more of the following components: valves, seats, cylinders, pistons, or gaskets.

Checking the Fuel Supply to the Carburetor

If the carburetor is equipped with a bowl drain, press the drain and allow a small amount of the fuel to leak from the bowl. If no drops appear, there may be an obstruction in the fuel tank, lines, or filter. If fuel does appear, check for trash and puddles of water. Either condition means that you should clean the fuel system. If a bowl drain is not available, you may need to loosen a fuel line fitting to see if fuel is reaching the carburetor. (See Figure 14-17.)

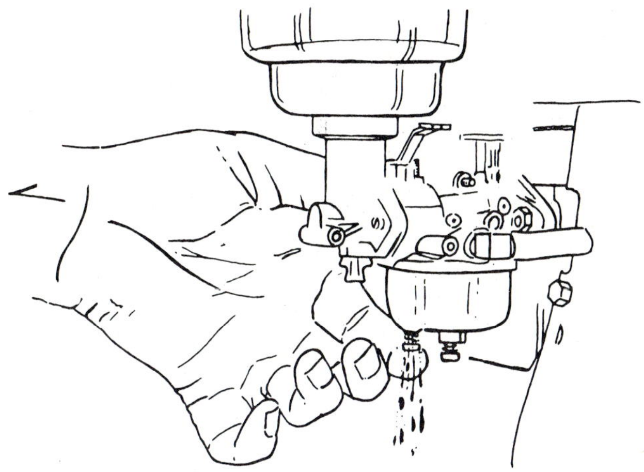

FIGURE 14-17. Checking for fuel in the bowl (Courtesy of Clinton Engines Corporation).

Spark Plug Condition

One of the most important steps in any troubleshooting procedure is to check the condition of the spark plug. The appearance of the plug is directly related to the condition of the engine. It is a visual history of how well (or poorly) the engine has been performing. The next figures and paragraphs describe how some typical engine conditions affect spark plug appearance.

Spark plugs that have been operated in a normal heat range and at varying engine speeds will have light brown or grey deposits. Unless there are signs of gross physical damage (like cracked insulators), such spark plugs may be cleaned, gapped, and reused. (See Figure 14-18.)

Spark plugs that are simply worn out from hours of use will have rounded electrodes and pitted insulators. They should be replaced. (See Figure 14-19.)

FIGURE 14-18. A normal spark plug (Courtesy of Clinton Engines Corporation).

FIGURE 14-19. A worn-out spark plug (Courtesy of Clinton Engines Corporation).

Dry, fluffy black deposits on the end of a spark plug indicate an excessively rich fuel mixture (too much gasoline). Some of the factors which might cause this condition are: (1) excessive choking, (2) clogged air cleaner, (3) poor ignition, and (4) prolonged idling at light loads (thus preventing the spark plug from getting hot enough to burn off the deposits). In the last case, using a spark plug with a higher heat range might help alleviate the problem. In any event, such "carbon fouled" spark plugs can often be cleaned and reused. (See Figure 14-20.)

Oil-fouled plugs have a wet oily appearance, often coupled with minor electrode erosion. This condition can result from oil slipping past the rings in new or recently rebuilt engines. However, if oil fouling occurs at any other time, it is a good sign that the engine may need rebuilding. (See Figure 14-21.)

Overheating is often indicated by spark plugs with badly burned or blistered electrodes. This can be caused by: (1) improper spark timing, (2) low octane fuel, (3) excessively rich fuel mixture, (4) cooling system obstructions or (5) sustained, high-speed, heavy-load operation. In the latter instance, using a lower heat range spark plug might help solve the problem. (See Figure 14-22.)

FIGURE 14-20. A carbon-fouled spark plug (Courtesy of Clinton Engines Corporation).

FIGURE 14-21. An oil-fouled spark plug (Courtesy of Clinton Engines Corporation).

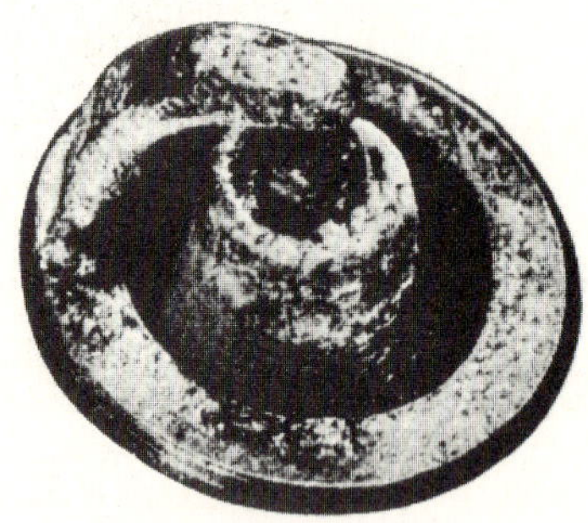

FIGURE 14-22. A spark plug with burned electrodes (Courtesy of Clinton Engines Corporation).

TUNE-UP

The extent of a tune-up will depend on the number and nature of the problems encountered during the troubleshooting operation. In the following discussion, we will limit ourselves to some of the most common types of adjustments and repairs.

Carburetor/Fuel System Service

Fuel system service can include a variety of tasks. Operations that don't involve the carburetor are usually straightforward – cleaning a filter, replacing a line, etc. Carburetor service may be more complex, ranging from a few adjustments to an overhaul.

Fuel system service is indicated when the other troubleshooting checks do not reveal a problem. If the fuel does not reach the carburetor, then the difficulty lies in a fuel line or filter. If fuel does reach the carburetor, the problem *might* be stale fuel in the tank. To be safe, replace old fuel. However, if everything else checks out okay, then the problem is likely to be the carburetor.

Adjusting the Carburetor

The first thing you should do is check the carburetor's external adjustment. A typical float bowl-type carburetor used on small engines has three external adjustment screws: (1) idle speed, (2) idle mixture, and (3) high-speed mixture. (See Figure 14-23.)

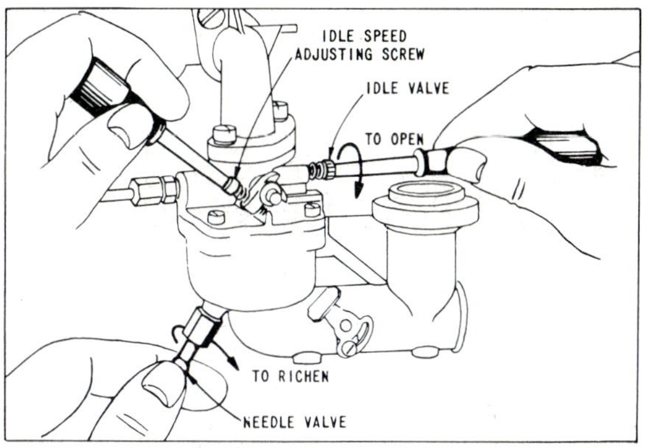

FIGURE 14-23. External adjustments to a float-type carburetor (Courtesy of Briggs & Stratton Corporation).

IDLE SPEED

Turning the idle screw in or out changes the closed position of the throttle plate and, hence, the idle speed of the engine. Although a specification (in revolutions per minute) is often provided by the manufacturer for idle speed, you can usually make the adjustment by "ear." In other words, turn the idle screw in or out until the engine is running as slowly as possible without stumbling or stalling.

IDLE MIXTURE

The idle mixture screw controls the proportion of air and fuel mixture at idle speeds. In most instances, the screw is associated with a needle valve.

When the screw is turned inward, the mixture is leaner, and when it is turned out, the mixture becomes richer. The mixture is adjusted while the engine is running at idle speed. First, turn the screw out until the engine begins to lose speed. Then turn the screw in until the engine just starts to run smoothly. Since this process may affect the idle speed, you will probably have to work back and forth between the idle speed and mixture screws until the most satisfactory settings have been obtained.

HIGH-SPEED MIXTURE

After the idle screws have been set, the high-speed mixture is adjusted. With the engine running under load, adjust the mixture screw until the engine will accelerate smoothly, without hesitation.

Rebuilding the Carburetor

If, after you have made all external checks and adjustments, the carburetor still seems at fault, it may need rebuilding. Here, in very general terms, are some of the major steps in rebuilding a typical, float bowl-type carburetor:

1. Before doing anything else, read the manufacturer's manual to become familiar with the components and operation of the carburetor you are working on. Follow the manufacturer's recommendations for detail procedures.
2. Remove the carburetor from the engine.
3. Disassemble the carburetor, being careful to lay out the pieces in the order they were removed.
4. Clean the carburetor components.
5. Inspect the needle valve and seat for wear. Replace if necessary.
6. Inspect the carburetor body and other components for wear and replace as required. (See Figure 14-24.)
7. Reassemble the carburetor up to the float bowl assembly.
8. Adjust the float bowl level according to the manufacturer's specifications. This is especially important. If the fuel level is too high, the mixture will be too rich. If the level is too low, the mixture will be too lean. (See Figure 14-25.)

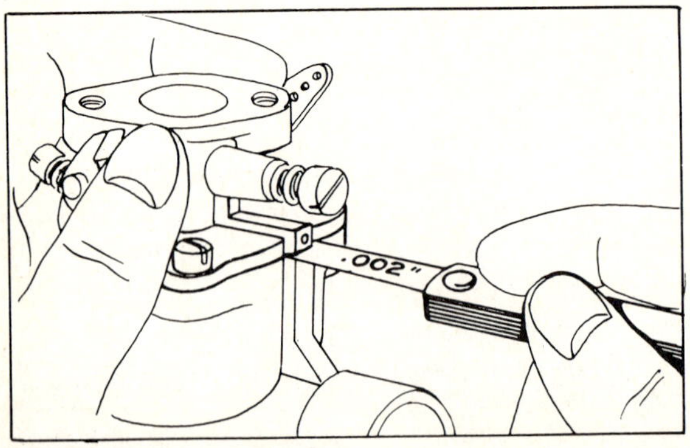

FIGURE 14-24. Checking a carburetor bowl cover for distortion (Courtesy of Briggs & Stratton Corporation).

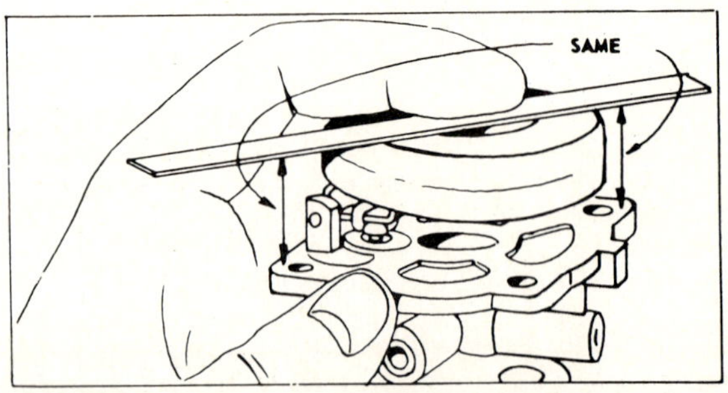

FIGURE 14-25. Checking the carburetor float level (Courtesy of Briggs & Stratton Corporation).

9. Reassemble the entire carburetor and install it on the engine.
10. Set the external adjustment screws to their approximately correct positions.
11. With the engine running, fine-tune the external adjustment screws.

IGNITION SYSTEM SERVICE

If the troubleshooting check leads you to suspect the ignition system, the first task is to eliminate the obvious sources of difficulty. Visually examine the spark plug wires for breaks or loose connections. Check the shutoff switch for shorts that might affect the ignition system.

If a quick visual inspection doesn't reveal a problem, then you probably need to service the points, condenser, or coil.

Preliminaries

Before any ignition system service can be performed, the components must be exposed. Following is a typical disassembly sequence:

1. Remove the air cooling/protective shroud from the flywheel. If the points are located outside the flywheel, most parts of the ignition system can now be serviced. However, in the majority of cases, the points are beneath the flywheel, which means that the next steps must be carried out.
2. Using a flywheel holder and wrench (to avoid damaging the flywheel fins), remove the nut that holds the flywheel on the crankshaft. (See Figure 14-26.)
3. Attach the correct puller and remove the flywheel from the crankshaft. (See Figure 14-27.)
4. Examine the key, the mating surface of the flywheel bore, and the end of the crankshaft. If the key has been sheared off, timing could be affected, which in turn could cause the engine to fail the "wide gap" plug test. If the key is intact, any ignition problems are likely to lie with the rest of the system. (See Figure 14-28.)

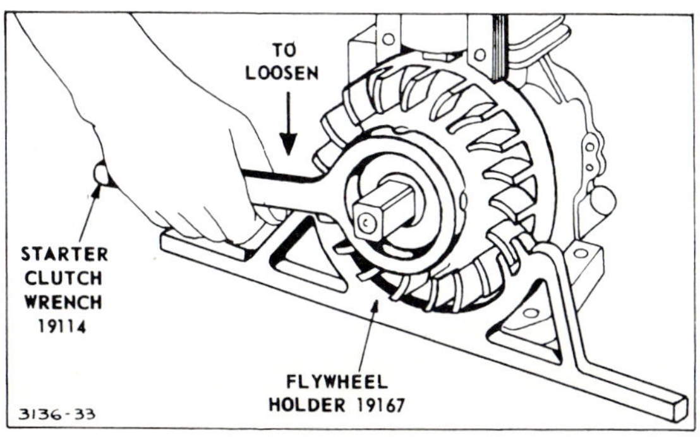

FIGURE 14-26. Loosening the flywheel (Courtesy of Briggs & Stratton Corporation).

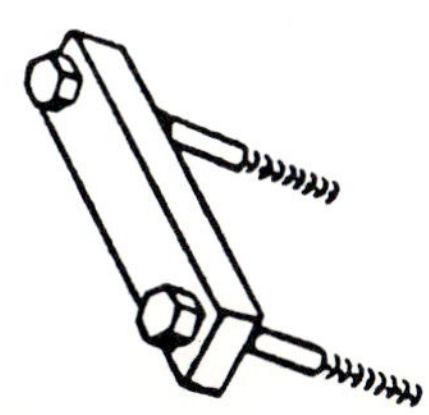

FIGURE 14-27. A flywheel puller (Courtesy of Briggs & Stratton Corporation).

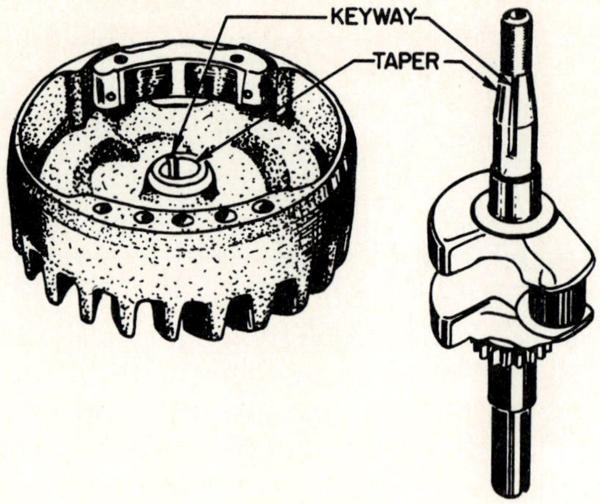

FIGURE 14-28. A flywheel and crankshaft (Courtesy of Clinton Engines Corporation).

Testing the Magnet

Before servicing the points and condenser, check the permanent flywheel magnet. If you don't have access to a magnetometer, place the tip of a screwdriver blade on the exposed magnet pole. If the blade sticks strongly, the magnet is probably okay.

Servicing the Points and Condenser

Usually, the points can now be exposed by removing a dust and grease cover. Two main types of points are likely to be encountered. In the example shown in Figure 14-29 the ignition wire is connected to the condenser, which itself forms the stationary point. Adjustments are performed by first setting the moving point at its highest position, then sliding the condenser point back and forth until the specified gap is obtained.

The other type of point assembly is shown in Figure 14-30. An eccentric adjustment is turned back and forth to regulate the point gap. When adjusting this type of point set, make sure that the points are parallel like those shown in Figure 14-31(a), and not skewed, like those pictured in Figure 14-31(b). It may be necessary to bend the point arm to achieve the correct alignment.

As you are adjusting any type of point assembly, avoid touching the points with your fingers. Engine oil on your hands or even your own natural oils can cause the points to burn on one side, which will shorten point life and reduce performance.

Old points may be adjusted if they are not visibly pitted, burned, or contaminated. Otherwise, they should be replaced. Whenever the points are replaced, a new condenser should also be installed. (See Figure 14-32.)

Coil Service

The coil has no moving parts and hardly ever wears out. However, heat is produced by current and may eventually break down the insulation. Since the primary and secondary windings are wrapped around the same frame, the entire coil must be replaced if either set of windings becomes defective.

The resistance to current flow offered by the coil can be checked with an ohmmeter. The results are then compared to the manufacturer's

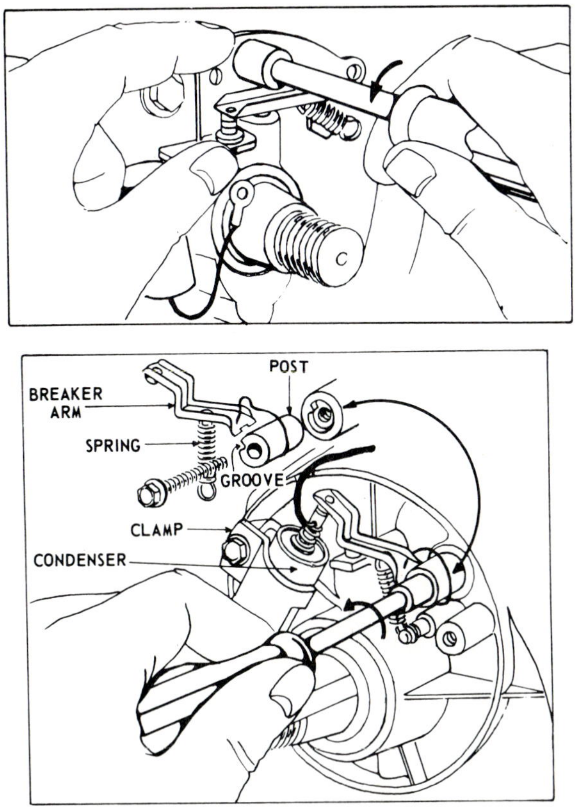

FIGURE 14-29. Breaker point assemblies (Courtesy of Briggs & Stratton Corporation).

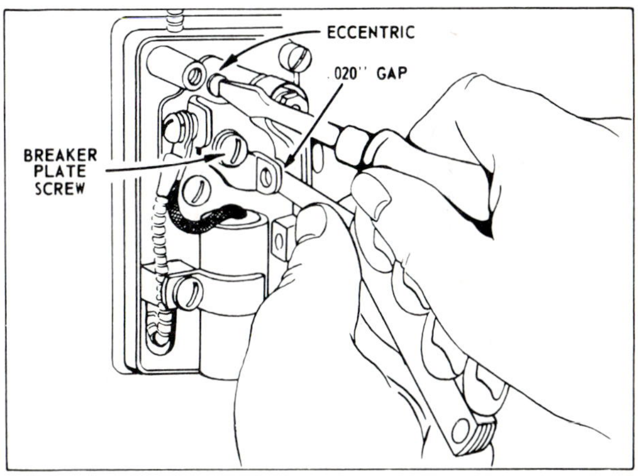

FIGURE 14-30. Adjusting breaker points (Courtesy of Briggs & Stratton Corporation).

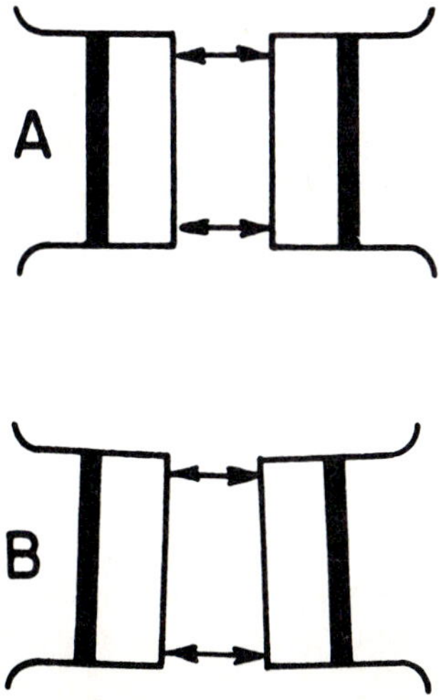

FIGURE 14-31. (A) Parallel points; (B) Skewed points (Courtesy of Clinton Engines Corporation).

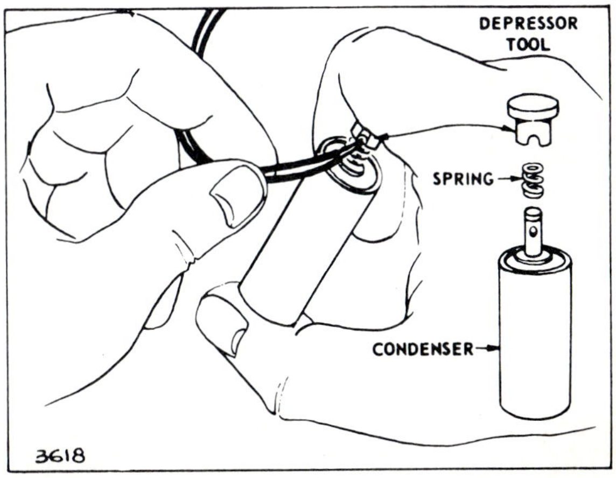

FIGURE 14-32. Assembling condenser and ignition wires (Courtesy of Briggs & Stratton Corporation).

specifications. To check the secondary (high-voltage) windings, remove the cable from the spark plug and place one ohmmeter lead on the end of the cable and the other lead on the coil frame. To check the primary (low-voltage) windings, remove the ignition wire from the points and connect one ohmmeter lead to the end of the wire and the other lead to the coil frame.

If a new coil is installed, make sure that the correct gap exists between the pole ends of the coil frame and the flywheel surface. Use a gauge made of paper or some other nonmagnetic material to set the gap. You will usually find that the coil frame is slotted to accommodate this type of adjustment. (See Figure 14-33.)

After replacing and/or adjusting the points and performing any other ignition system service, replace the dust cover and carefully install the flywheel and key. Put the shroud back on the engine. If the spark plug has been removed (which is likely), set the electrode to the correct gap before screwing the spark plug in place. (See Figure 14-34.)

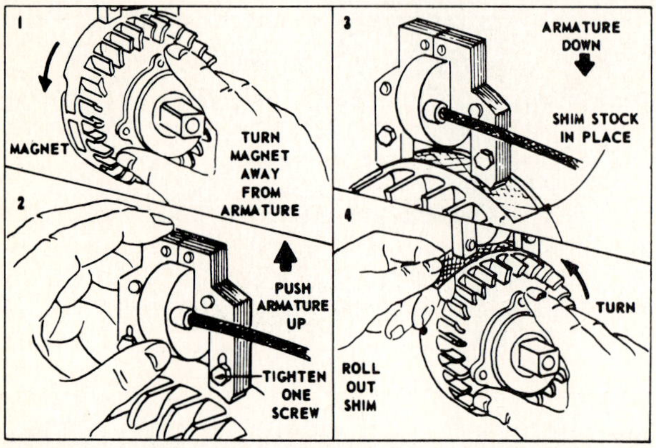

FIGURE 14-33. Adjusting coil armature air gap (Courtesy of Briggs & Stratton Corporation).

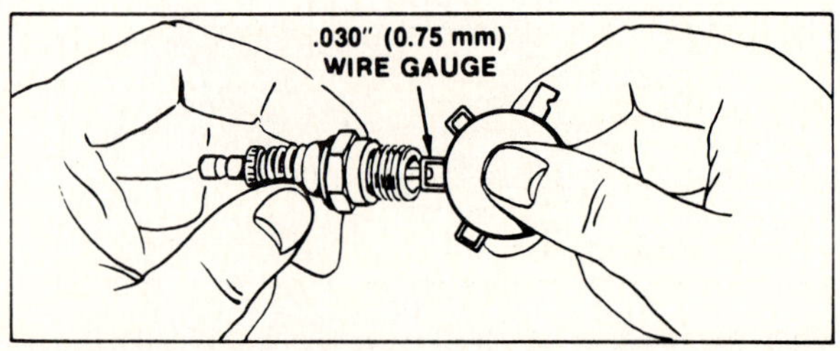

FIGURE 14-34. Adjusting spark plug electrode gap (Courtesy of Briggs & Stratton Corporation).

Timing

Ignition timing is not adjustable on most small engines. The exact moment of ignition is permanently set at the time of manufacture. However, some engines have a magnetic rotor behind the flywheel (as opposed to a magnet embedded in the surface of the flywheel). The timing of these engines is adjusted by moving the position of the magnetic rotor. (See Figure 14-35.)

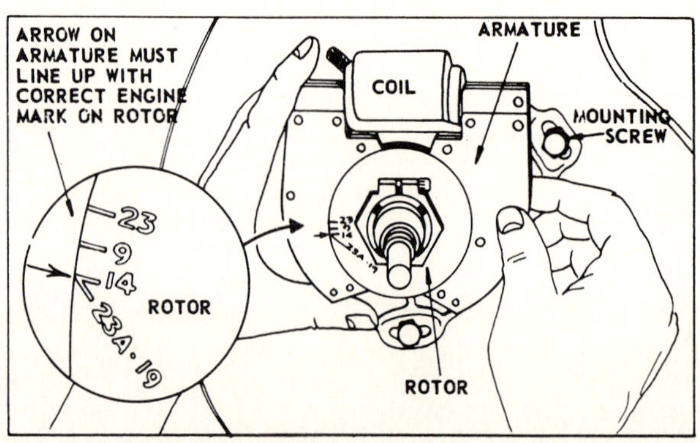

FIGURE 14-35. Adjusting rotor timing (Courtesy of Briggs & Stratton Corporation).

MANUAL STARTER REPAIR

Although it is not considered part of a regular tune-up, you may be called on from time to time to fix a recoil rope starter. Here is a typical repair sequence:

Removing a Starter Spring

1. Cut the rope knot at the starter pulley.
2. Pull the rope out of the pulley.
3. Grasping the outer end of the recoil spring, pull it out of the starter housing as far as possible. (See Figure 14-36.)
4. Bend one of the bumper tangs up and lift out the starter pulley, thereby disconnecting the spring. (See Figure 14-37.)

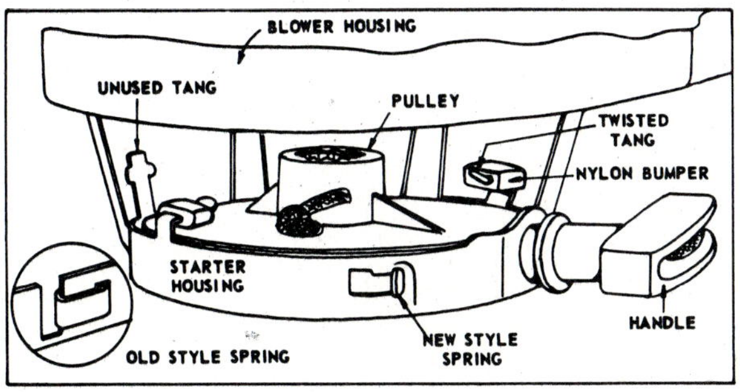

FIGURE 14-36. A starter assembly (Courtesy of Briggs & Stratton Corporation).

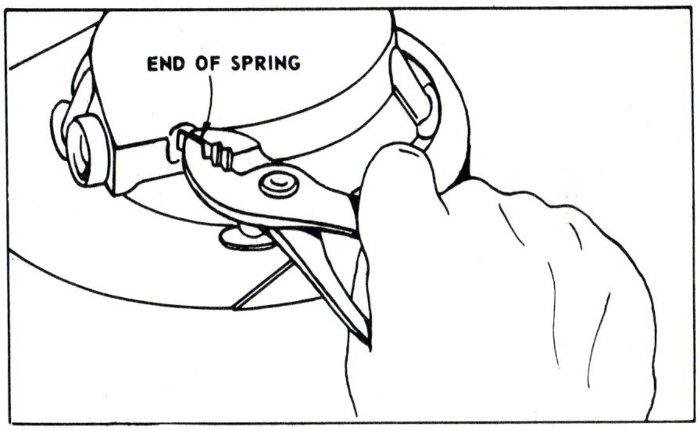

FIGURE 14-37. Removing a rope rewind spring (Courtesy of Briggs & Stratton Corporation).

Installing a Starter Spring

5. If reinstalling an old spring, clean it with solvent. Wipe the spring dry by pulling it through a cloth.
6. Insert either end of the spring into the blower housing slot. Hook the other end on the pulley. Replace the nylon bumpers if they are worn.
7. Place a dab of grease on the pulley; set the pulley into the housing and bend the bumper tang down. (See Figure 14-38.)

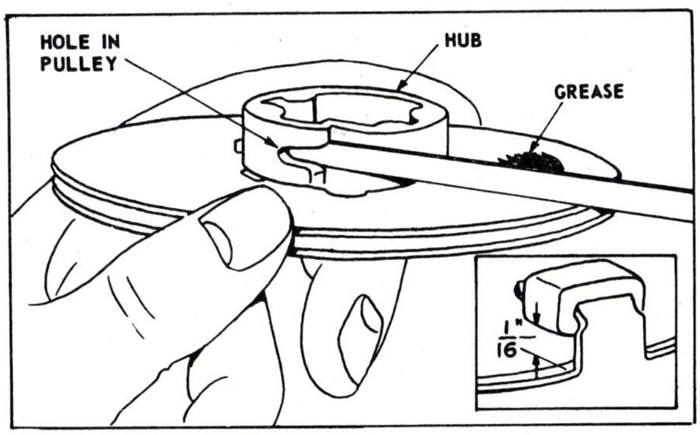

FIGURE 14-38. Installing a starter rewind spring (Courtesy of Briggs & Stratton Corporation).

Winding the Spring

8. Place a ¾ in. piece of square stock into the center of the pulley hub.
9. Grasping the stock with a wrench, wind the pulley until the spring is tight.
10. Back off the pulley one turn—until the rope knot hole in the pulley and the eyelet in the blower housing line up.

Installing a New Rope

11. Check the specifications for rope size and length.
12. Insert the rope through the handle and tie a figure eight knot in the end of the rope.
13. Insert a pin through the figure eight and pull the knot and pin tightly into the handle. (See Figure 14-39.)
14. If necessary, fashion a rope inserter tool from a piece of music or spring wire. An example is shown in Fig. 14-40.
15. Thread the rope and inserter wire through the rope eyelet hole in the housing and out the pulley hole. (See Figures 14-41 and 14-42.) *Caution*: The rope must pass through a guide lug in old-style pulleys.
16. Tie a knot in the rope, and then pull the knot tight. Work the knot down into the pulley hole. (See Figure 14-43.)

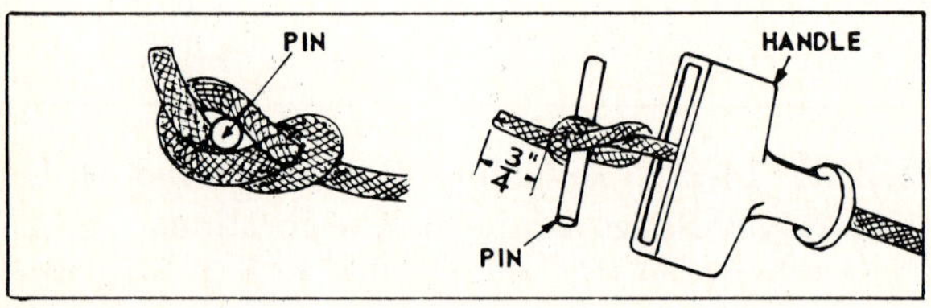

FIGURE 14-39. Installing a pull rope (Courtesy of Briggs & Stratton Corporation).

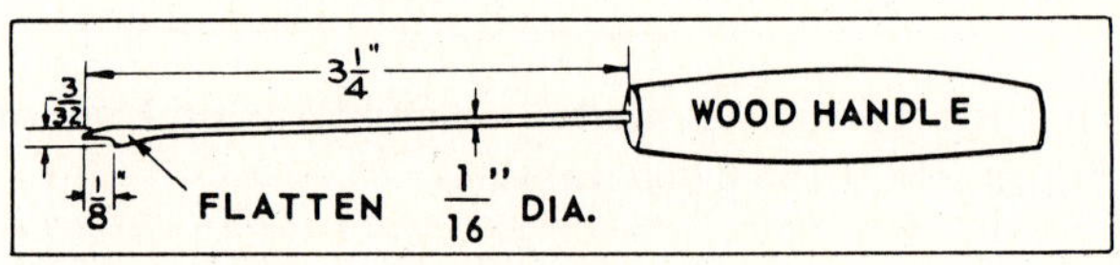

FIGURE 14-40. A pull rope inserter (Courtesy of Briggs & Stratton Corporation).

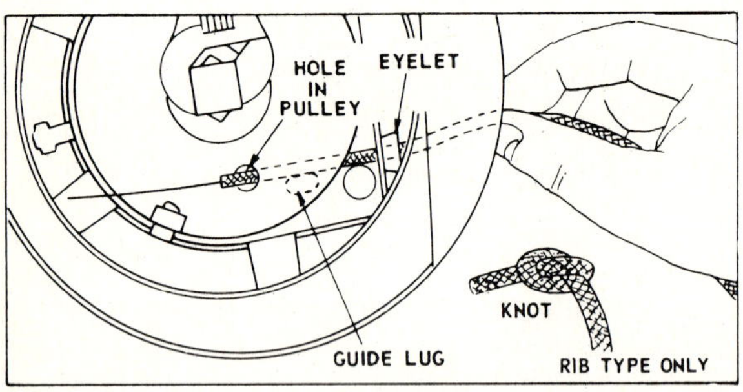

FIGURE 14-41. Inserting rope (Courtesy of Briggs & Stratton Corporation).

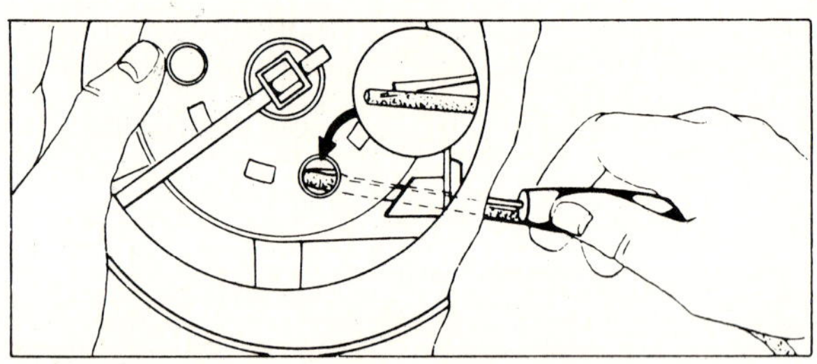

FIGURE 14-42. Inserting a pull rope (Courtesy of Clinton Engines Corporation).

ELECTRICAL TESTING AND REPAIR

The remainder of this chapter deals with specialized electrical service. However, it should be noted that for most small engines such service will

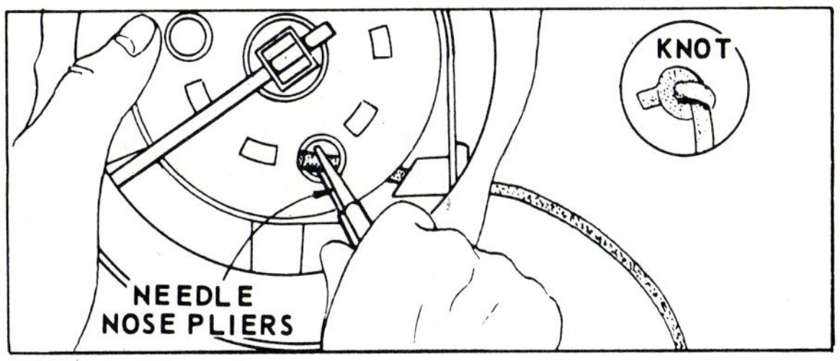

FIGURE 14-43. Tying a knot (Courtesy of Briggs & Stratton Corporation).

not be necessary. This is because most engines do not include any electrical components (other than the ignition system). Also, even if the electrical components are present, the work may be farmed out to electrical specialty shops.

Therefore, the following topics will only be covered in general, introductory terms.

Electrical Test Equipment

It is not often (or, very rarely) possible to completely determine the condition of an electrical device or circuit by a visual inspection. Usually, electrical test equipment is required. Here, briefly, are some common types of electrical test gear:

MULTIMETER

Contains a voltmeter, ammeter, and ohmmeter in one case. Controls are provided to select the desired test function as well as the measurement range or scale. The meter contains a pair of test leads. The ends of the leads may be simple probes, or they may be alligator clips. (See Figure 14-44.)

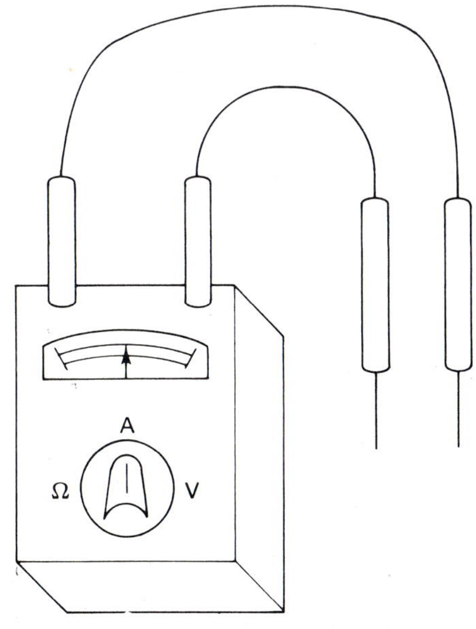

FIGURE 14-44. A multimeter.

AMMETERS

A separate ammeter may be required if a circuit carries more current than a multimeter is capable of reading. (See Figure 14-45.)

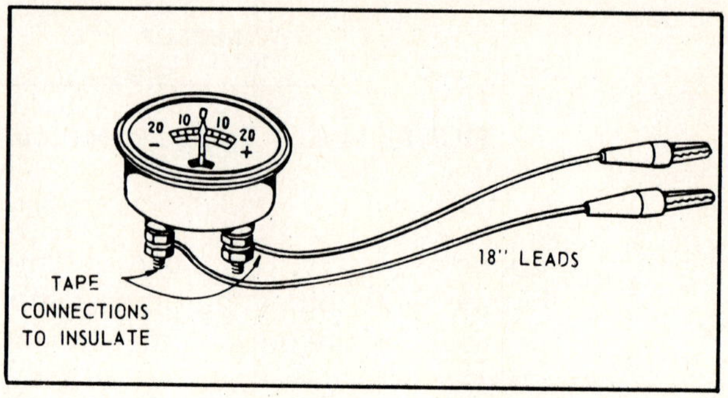

FIGURE 14-45. A test ammeter (Courtesy of Briggs & Stratton Corporation).

TEST LIGHT

A test light is nothing more than two leads connected to a light bulb. This piece of test gear, which can be purchased or assembled on your own, is used to determine whether current is flowing through a circuit. (See Figures 14-46 and 14-47.)

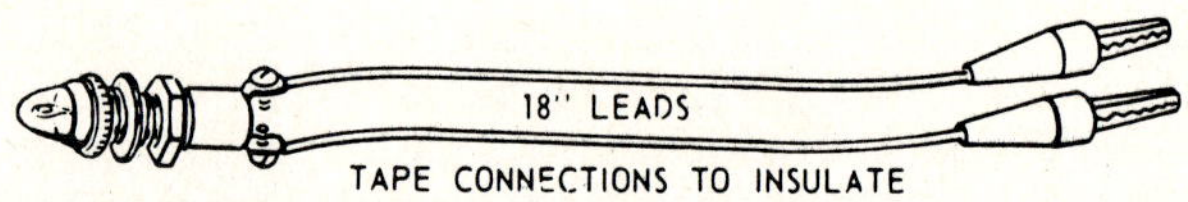

FIGURE 14-46. A test lamp (Courtesy of Briggs & Stratton Corporation).

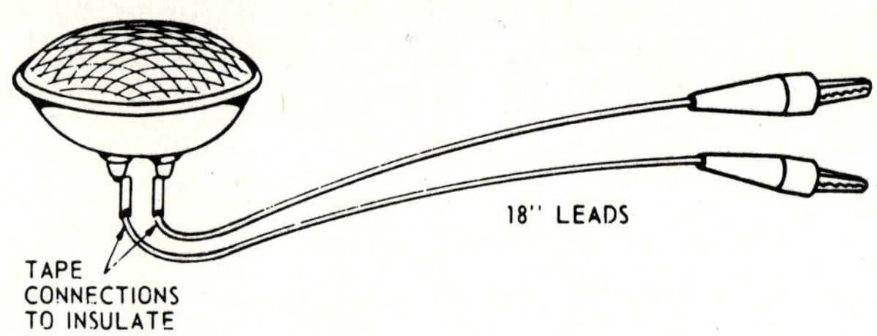

FIGURE 14-47. A load test lamp (Courtesy of Briggs & Stratton Corporation).

TEST LEAD

Used to bypass (or "short out") a section of circuit. The example shown in Figure 14-48 has an alligator clip at one end and a fuse connector on the other.

Sample Electrical Tests

The following paragraphs and figures illustrate some common test procedures.

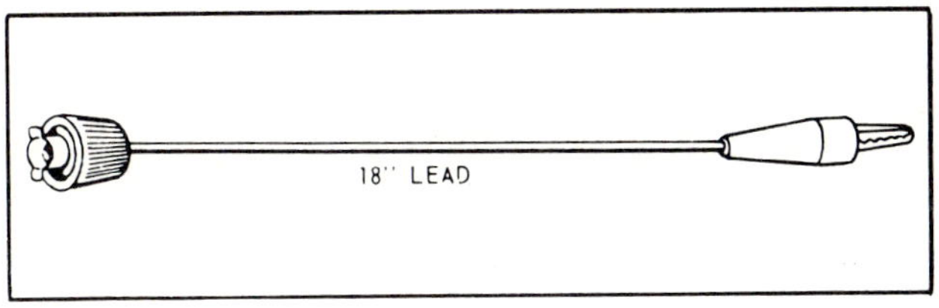

FIGURE 14-48. Fuse cap test lead (Courtesy of Briggs & Stratton Corporation).

CHARGING SYSTEM

1. Figure 14-49 shows a small engine headlight circuit being checked with a test light made from a headlight bulb. Naturally, if the bulb shines, the circuit is okay.
2. If the small engine electrical system includes a storage battery, then the charger must produce enough current to charge the battery. This test shows the ammeter hookup required to test the charger in one type of system. (See Figure 14-50.)
3. The test shown in Figure 14-51 pictures a voltmeter being used to check the voltage available in a system which employs a NICAD battery. The reading is approximately 13.5 volts, which is correct for this type of system. Although a voltmeter is illustrated, a multimeter could be used—so long as the proper function and scale are selected.
4. Figure 14-52 shows a multimeter being used to check the diode rectifier assembly in an AC charging system. To perform this test, the battery is first disconnected from the system, and then the meter is turned to the resistance (ohmmeter) function. When set in this position, the meter will send a known current through the circuit being tested. Ohms are an indication of the resistance to current flow. When the diode assembly is working properly, it will allow current to flow in one direction, but not in the other. Therefore, if the meter leads are connected as shown, then swapped, the needle should fluctuate from higher to lower resistance readings (or vice versa). If the reading stays the same (either high or low), the rectifier assembly is not working. *Caution*: When using an

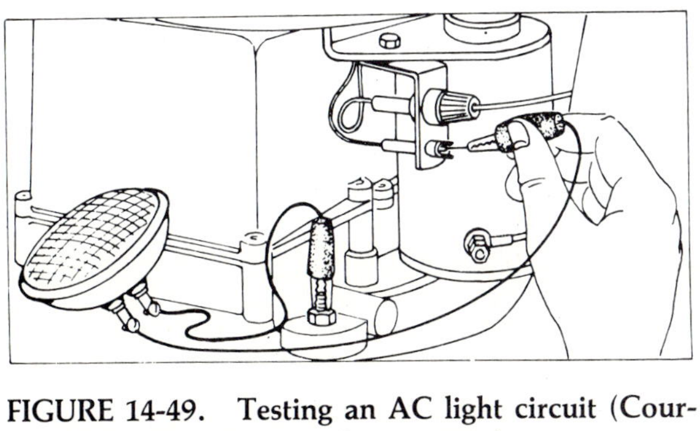

FIGURE 14-49. Testing an AC light circuit (Courtesy of Briggs & Stratton Corporation).

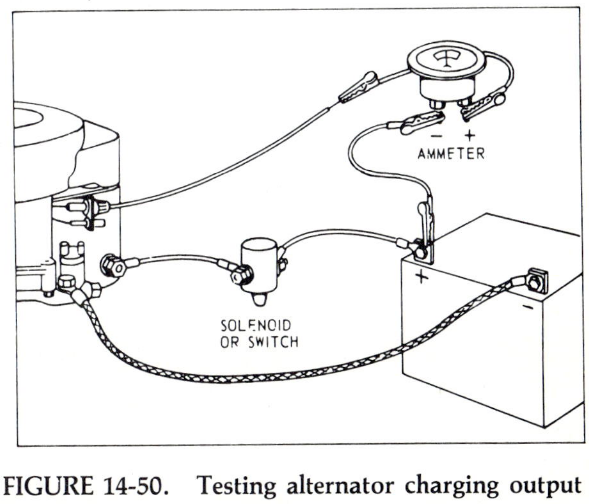

FIGURE 14-50. Testing alternator charging output (Courtesy of Briggs & Stratton Corporation).

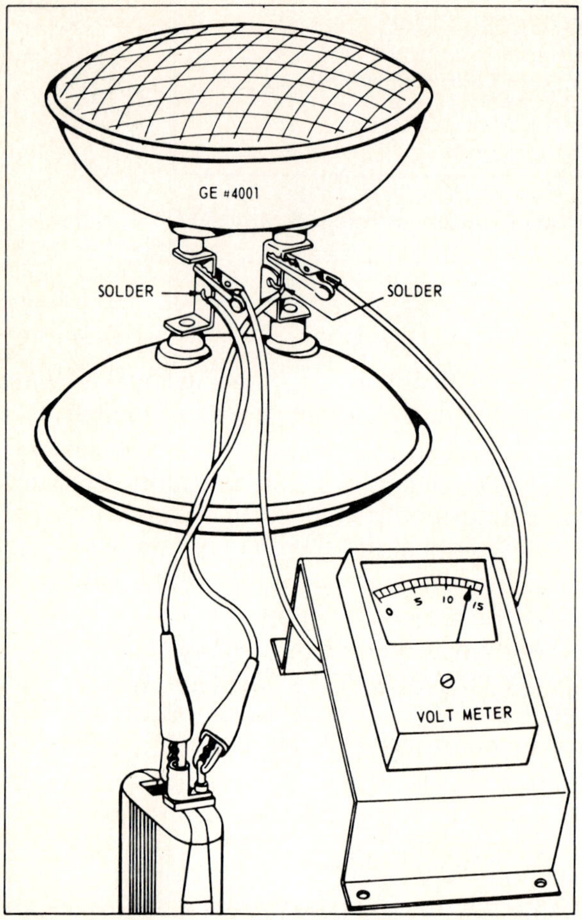

FIGURE 14-51. A battery tester (Courtesy of Briggs & Stratton Corporation).

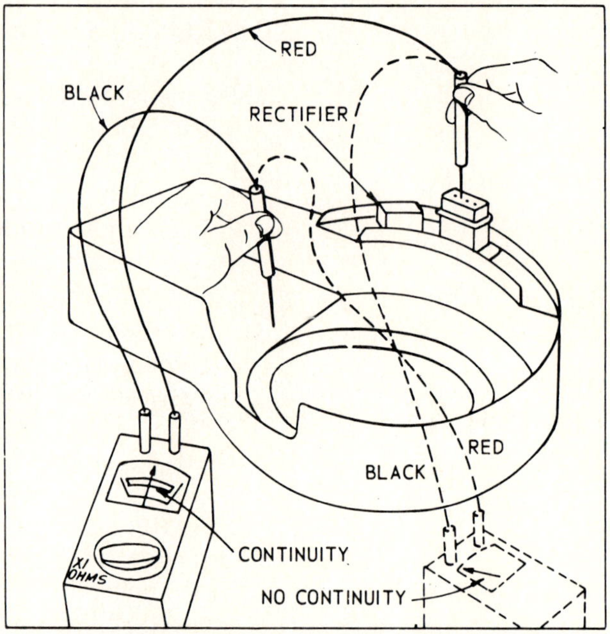

FIGURE 14-52. Testing an alternator rectifier (Courtesy of Briggs & Stratton Corporation).

ohmmeter, always make sure that no current from an outside source enters the meter. Otherwise, damage to the meter may result.

5. This test (Figure 14-53) pictures a voltmeter being used to check the output of an alternator charging system. The test is performed by disconnecting the charge lead, then connecting the voltmeter between the end of the charge lead and ground post of the battery. If the system is functioning properly, the reading should be about 14 volts. When this test is performed, special care must be taken to connect the negative test lead to the negative side of the circuit and the positive lead to the positive side. Otherwise, the meter may be damaged.
6. Figure 14-54 shows a continuity check being performed to determine if a circuit is complete. In this particular example, the test will determine if a problem lies in the wiring harness or in the alternator windings.

ELECTRIC STARTER TESTING (Figure 14-55)

7. You may on occasion need to check the brushes in an electric starter motor (Figure 14-55). In the example shown in Figure 14-56, the motor is disassembled by removing the "through" bolts. Once the brushes have been removed, their length should be compared to the length of

new brushes. If the length is one-half or less than the original size, the brushes should be replaced. Also, the brush spring tension should be checked against the factory specification. (See Figure 14-57.)

8. Starter motors can be mounted on a bench vise to check speed and current draw. Figure 14-58 shows an ammeter and a tachometer connected to a bench-mounted starter motor.
9. Figure 14-59 pictures the hookups for checking a 12-volt starter system.
10. Figure 14-60 shows how current draw and speed are checked on a 110-volt motor.

Caution: Be especially careful when checking 110-volt motors. Electrical shocks from higher-voltage systems can be lethal.

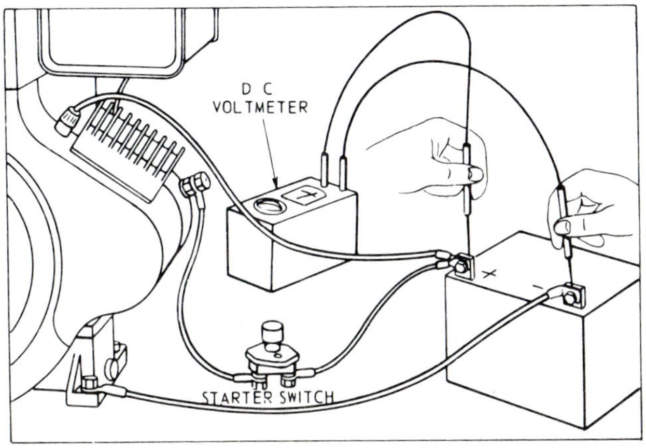

FIGURE 14-53. Rectifier regulator check (Courtesy of Briggs & Stratton Corporation).

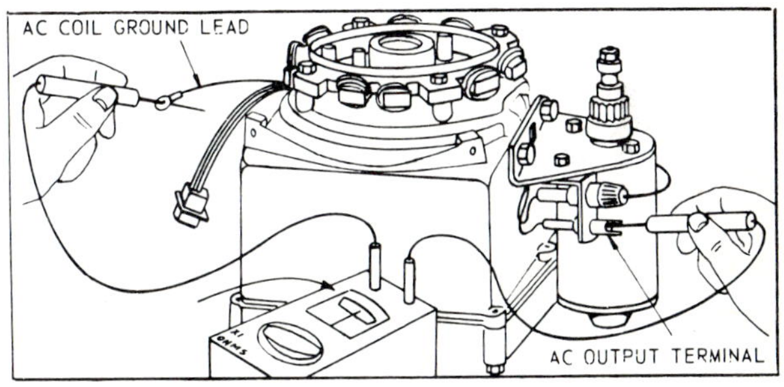

FIGURE 14-54. Testing AC light circuit output (Courtesy of Briggs & Stratton Corporation).

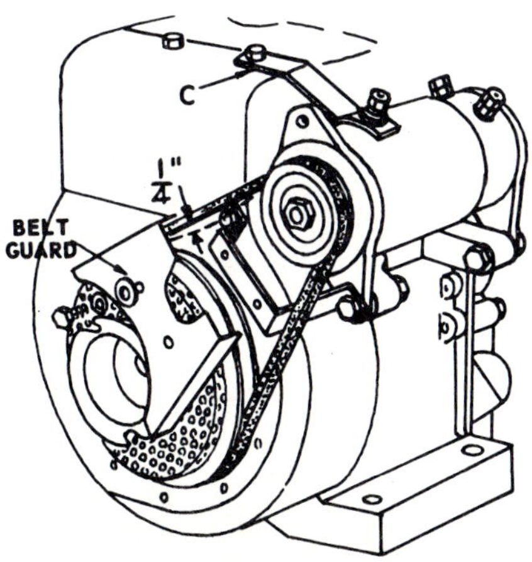

FIGURE 14-55. A starter generator (Courtesy of Briggs & Stratton Corporation).

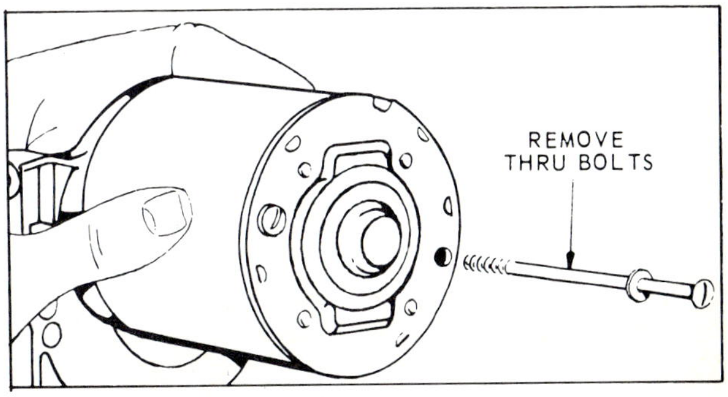

FIGURE 14-56. Removing "thru" bolts (Courtesy of Briggs & Stratton Corporation).

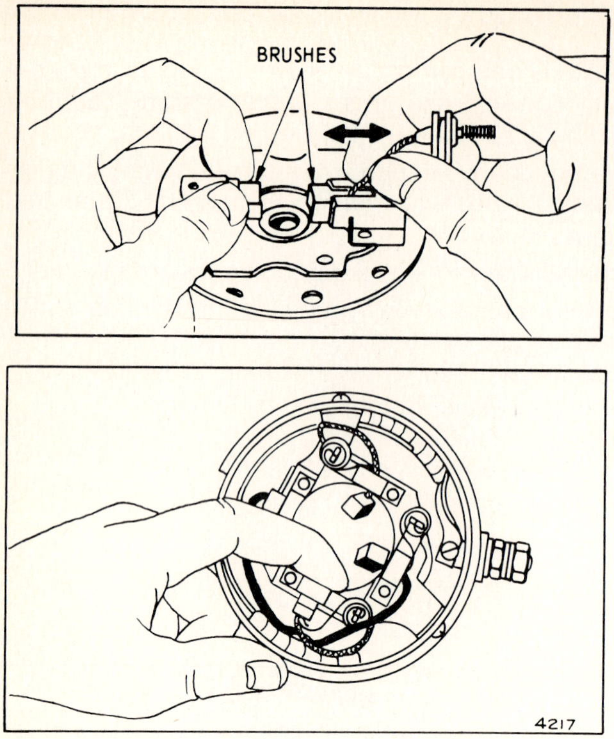

FIGURE 14-57. Checking brushes (Courtesy of Briggs & Stratton Corporation).

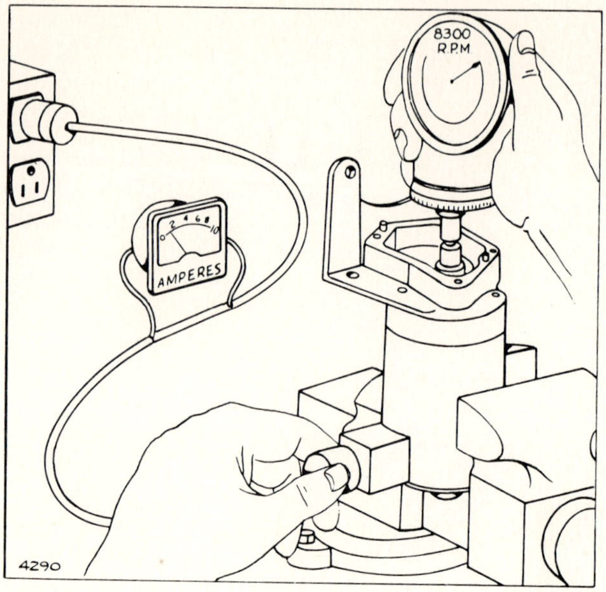

FIGURE 14-58. Checking 110-volt starter performance (Courtesy of Briggs & Stratton Corporation).

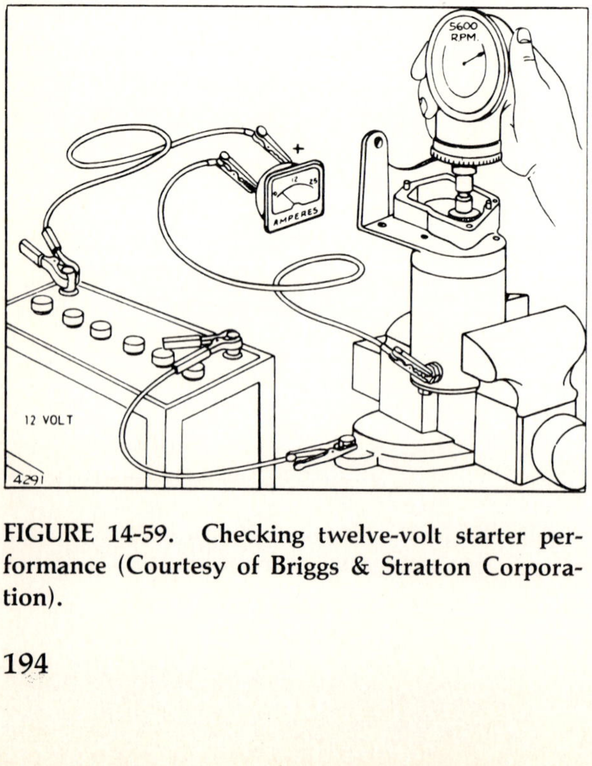

FIGURE 14-59. Checking twelve-volt starter performance (Courtesy of Briggs & Stratton Corporation).

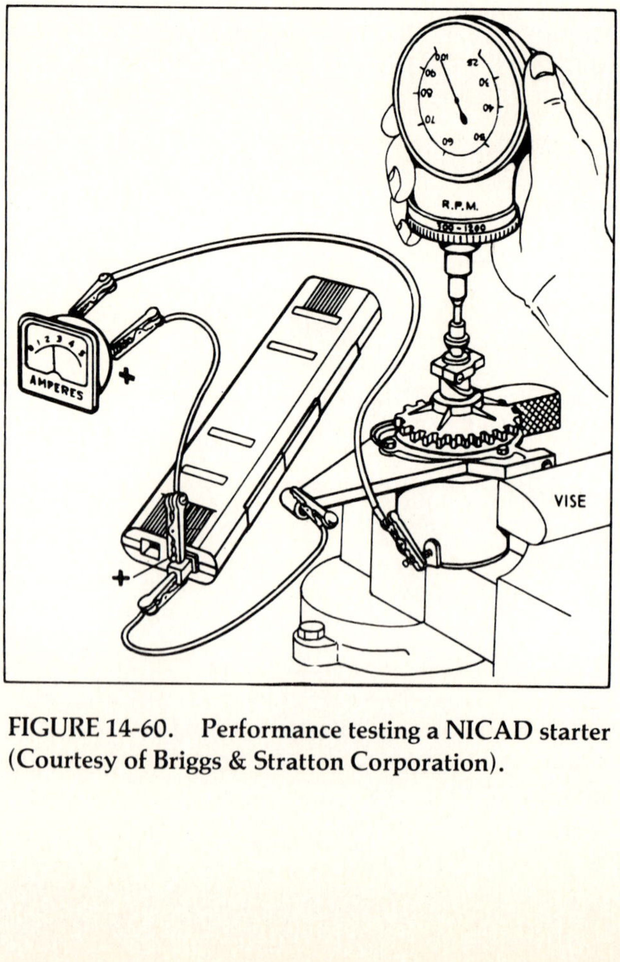

FIGURE 14-60. Performance testing a NICAD starter (Courtesy of Briggs & Stratton Corporation).

15

HAND TOOLS

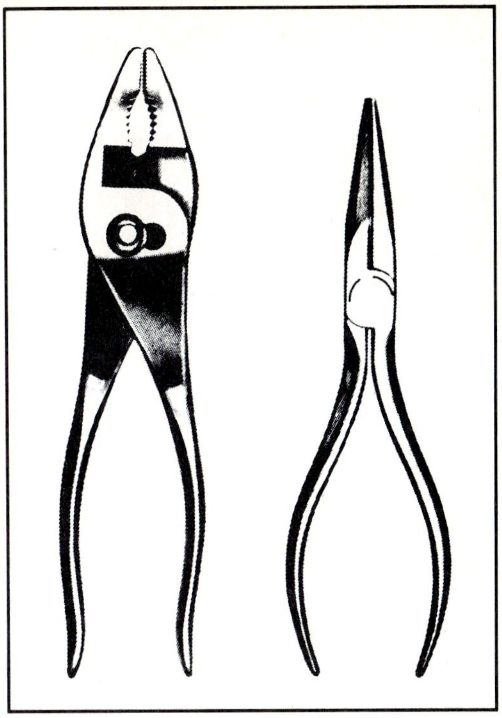

GENERAL

Hand tools, as the name implies, are devices that use human muscle power for operation. Special tools can provide gripping, turning, or cutting power not available to the unassisted body. This chapter is about the common hand tools used for small gasoline engine repairs.

SOME REMINDERS

Before reviewing the common hand tools and their functions, it will be helpful to remember some basic facts about tool use.

1. Although everybody repeats this statement, it is still true: Whenever possible, use the correct tool for the job. The wrong tool slows you down and in some cases can cause you to botch a job. For certain tasks, having the right tool is absolutely necessary. For instance, you shouldn't even attempt to remove a cylinder head unless you have a torque wrench to put it back on.
2. Never use any more force than is necessary when operating a tool. Anyone who has ever twisted a bolt head off doesn't need to be reminded of this rule.
3. Identify your own tools and be careful about either borrowing or lending tools. Since professional mechanics are expected to supply their own hand tools, they are apt to be sensitive about the loss of or damage to expensive equipment.
4. Although some mechanics seem to perform very well with a sloppy tool chest full of dirty tools, it is probably better, at least starting out, to strive for neatness.

5. Although the most expensive, chrome-plated tools may not be worth the money, neither are the cheapest tools. If you are a professional, buy equipment that will last a long time. If you are an amateur, you still ought to get the best tools you can afford. They will help make up for your shortcomings in other areas.

A TYPICAL TOOL BOX

The following paragraphs list the items commonly found in a well-equipped tool box.

Screwdrivers

The two most common types of screwdrivers are the Phillips and the slotted-blade kind. They come in all types and sizes; make sure to use big screwdrivers for big screws and little ones for little screws. This may seem obvious, but from an examination of the random methods many people seem to employ when selecting a screwdriver, it really isn't. (See Figures 15-1 and 15-2.)

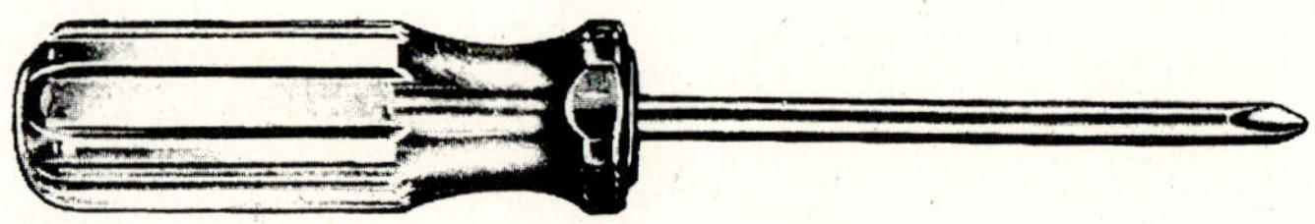

FIGURE 15-1. A Phillips screwdriver with a plastic handle (Courtesy of New Britain Hand Tools).

FIGURE 15-2. A slotted blade or common bit screwdriver (Courtesy of New Britain Hand Tools).

Nut Drivers

A nut driver looks like a screwdriver with a socket wrench fitting at the end. Various (small) size sockets can be connected to quickly attach or remove small nuts. (See Figure 15-3.)

FIGURE 15-3. A nut driver (Courtesy of New Britain Hand Tools).

Pliers

Various kinds of pliers are available: Standard gripping pliers, needle-nose pliers, diagonal cutting pliers, bolt-cutting pliers, channel lock pliers, etc. If

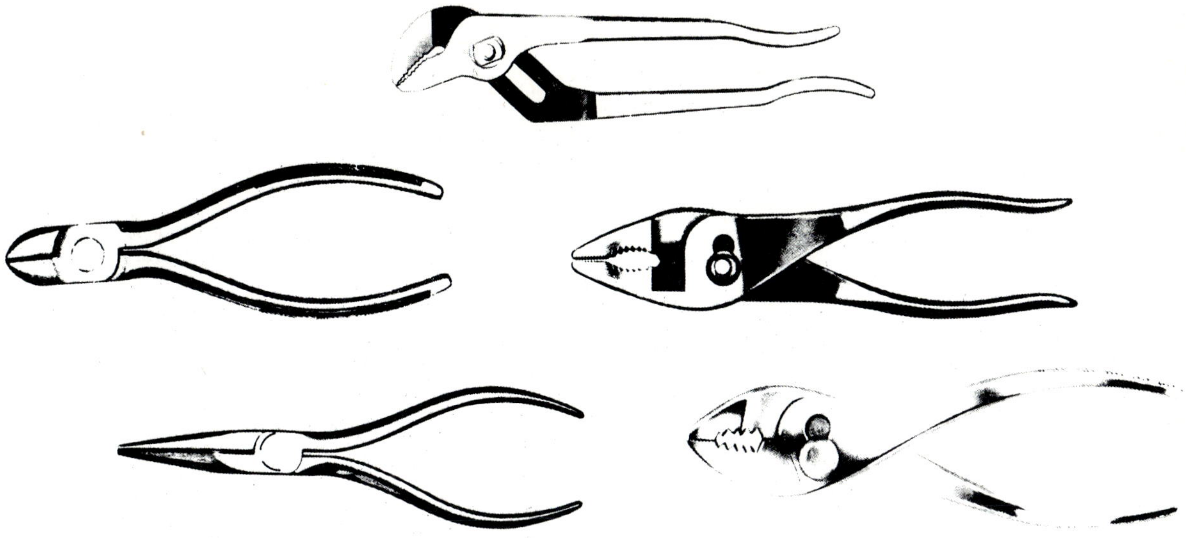

FIGURE 15-4. **Pliers (Courtesy of New Britain Hand Tools).**

the proper type of pliers is going to be used, it is not a bad idea to have one each of the major types. (See Figure 15-4.)

Hammers

Two types of hammers are commonly found in a mechanic's tool box: A steel, ball peen hammer, and a soft face mallet of some kind. (See Figure 15-5.) The ball peen should only be used as a last resort. Its indiscriminate use causes more problems than cures.

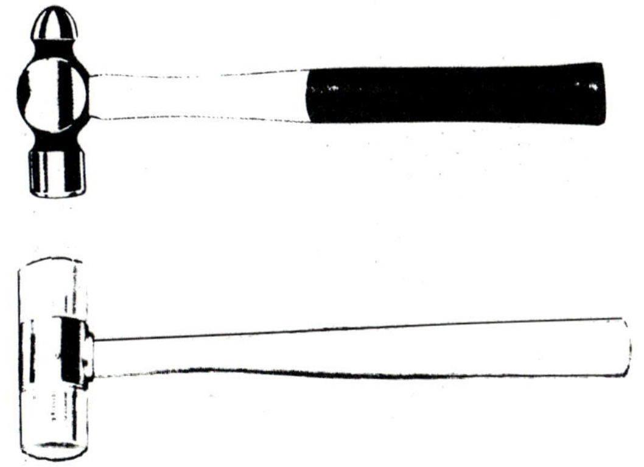

FIGURE 15-5. **A ball peen hammer and a soft-face hammer (Courtesy of New Britain Hand Tools).**

Wrenches

Three types of wrenches are commonly used: (1) open end, (2) box end, and (3) combination open/box end wrenches. Each performs the same function; however, the open end is easier to get on and off the nut, whereas the box end can sometimes get into a tighter space; it also has more gripping power. A well-equipped tool box should have a complete range of each kind of wrench. (See Figures 15-6, 15-7, and 15-8.)

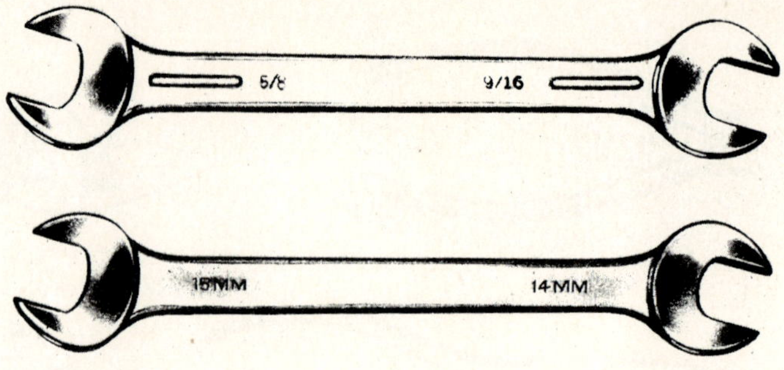

FIGURE 15-6. A standard open-end wrench and a metric open-end wrench (Courtesy of New Britain Hand Tools).

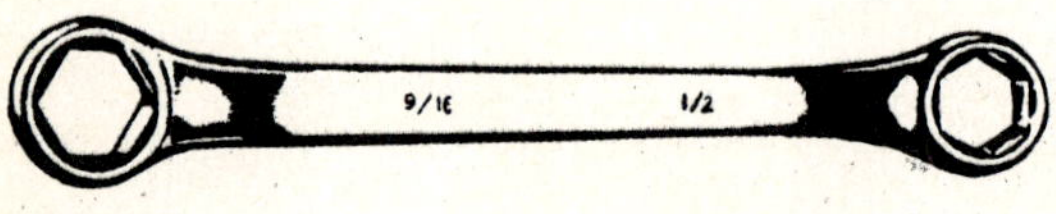

FIGURE 15-7. A six-point box-end wrench (Courtesy of New Britain Hand Tools).

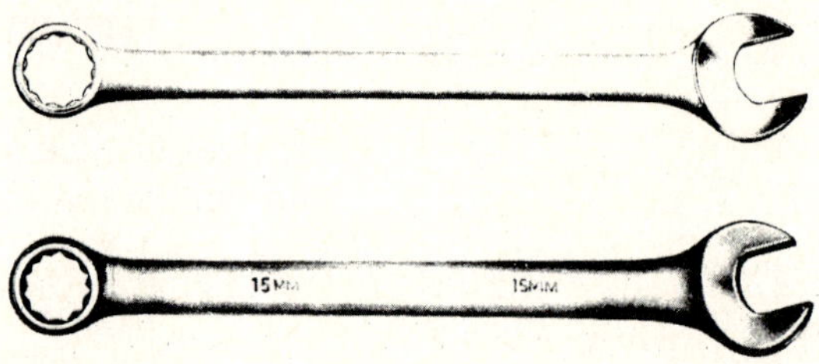

FIGURE 15-8. Combination wrenches (Courtesy of New Britain Hand Tools).

Socket Wrenches

Any serious mechanic, whether professional or amateur, should have a socket wrench set with a complete set of sockets and attachments. Work simply goes too slowly without this equipment. (See Figure 15-9.)

Adjustable Wrenches

Adjustable wrenches are handy but don't provide the positive grip necessary for most work. Improperly used adjustable wrenches can quickly round off a nut or bolt head. (See Figure 15-10.)

Impact Screwdrivers

Impact screwdrivers are used to loosen stubborn fasteners. The tool is held in one hand while the end is struck with a hammer wielded by the other hand. The force of the hammer impact is translated by a cam action into turning power. Different bits (standard screwdrivers, Phillips, etc.) are available. There are occasions when an impact screwdriver is the only tool that can do the job.

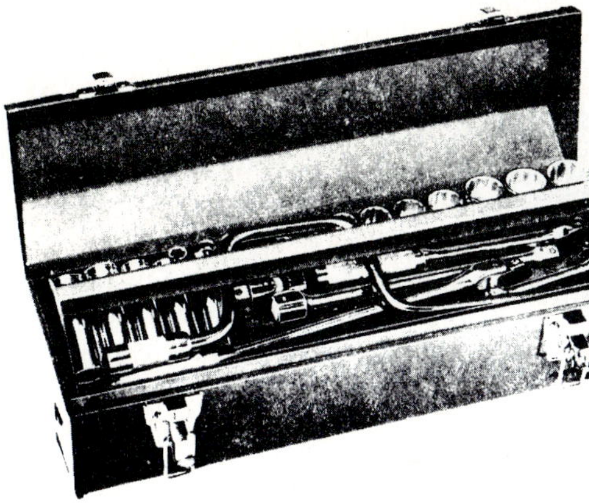

FIGURE 15-9. A socket wrench set (Courtesy of New Britain Hand Tools).

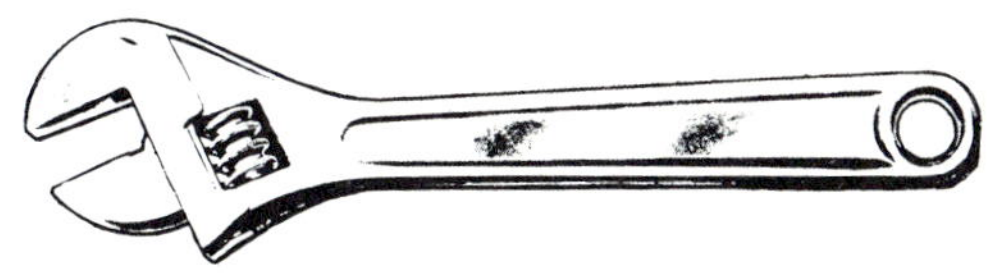

FIGURE 15-10. An adjustable wrench (Courtesy of New Britain Hand Tools).

Hand-held Cutting Tools

Mechanics often need to cut or shape metal in various ways. A complete tool box should include:

1. A hacksaw (Figure 15-11)
2. A set of chisels and punches (Figure 15-12)
3. A set of files (Figure 15-13)
4. A tap and die set for cutting threads (Figure 15-14)

FIGURE 15-11. A hacksaw (Courtesy of New Britain Hand Tools).

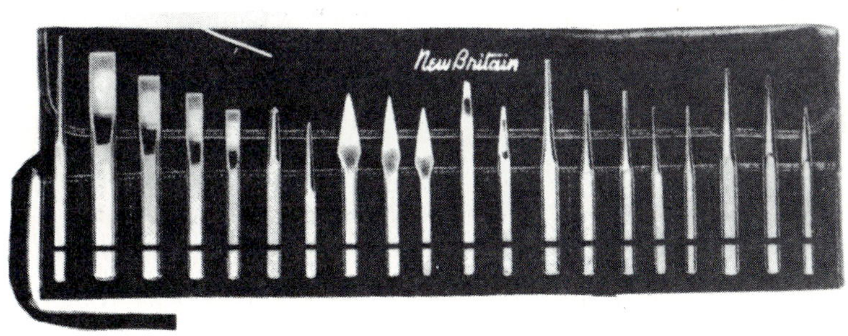

FIGURE 15-12. A chisel and punch set (Courtesy of New Britain Hand Tools).

FIGURE 15-13. A mill bastard file and a round bastard file (Courtesy of New Britain Hand Tools).

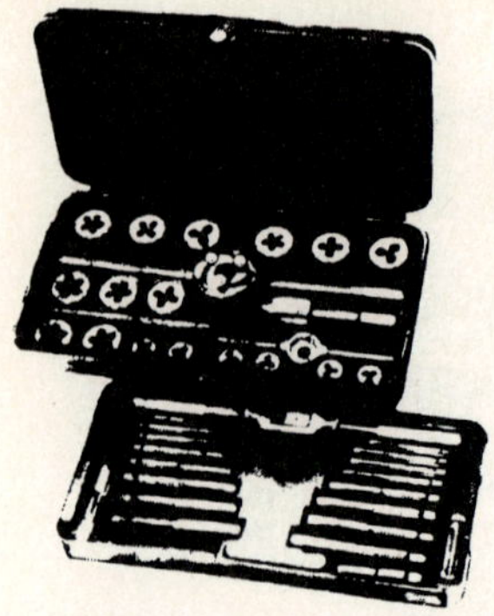

FIGURE 15-14. A mechanic's tap and die set (Courtesy of New Britain Hand Tools).

SPECIAL TOOLS

There are many special tools that make a mechanic's job easier. Some are special tools produced by manufacturers for specific kinds of engines. These tools aren't usually bought by the individual mechanic unless he is working in a very small shop. Many mechanics custom-tailor existing tools to suit their own preferences.

16

MEASURING TOOLS

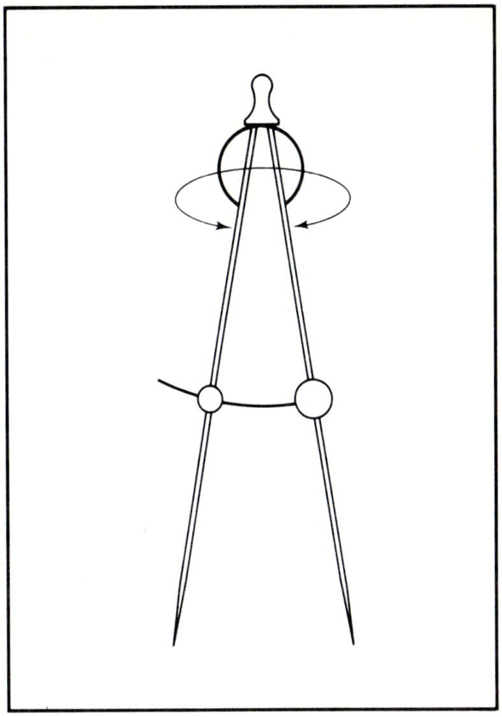

GENERAL

Besides needing tools to increase the gripping, turning, or cutting power of the human hand, tools are also required to enhance the judgment of the eye and brain. We must be able to make accurate comparisons between different sizes of things and spaces. We also must be able to say accurately just how large an object or space is or how much something weighs. This is particularly important in engine work because a few thousandths of an inch or a few grams' difference between two measurements can have a serious effect on engine performance.

The following paragraphs list some common measuring tools that might be used by mechanics who work on small gasoline-powered engines.

COMMON TAPE MEASURES AND STRAIGHT RULERS

Most tool boxes contain one or both of these measuring devices. They are adequate for making straight measurements of longer distances. However, most engine-related measurements are less than six inches, often less than an inch. Also, many engine measurements are not made on flat surfaces along straight lines. The distance along curved surfaces is often measured. Furthermore, it is often necessary to measure the void or gap between two surfaces. Consequently, more specialized measuring tools are needed for engine work.

FEELER GAUGES

Feeler gauges are used to measure gaps. Three types of feeler gauges are commonly used.

1. *Blade type* feeler gauges consist of a set of precisely made flat blades, each of which is a specified thickness. Different size blades are inserted into the gap being measured until one fits. The size of the blade that gives the best fit corresponds to the size of the gap. (See Figure 16-1.)
2. *Wire type* feeler gauges are made along the same lines as a blade-type feeler gauge. However, instead of blades, calibrated wires are used.
3. *Go, no-go blade type* feeler gauges consist of blades thicker at one end than another. This type of feeler gauge is often used where the specification for a gap is given as a high-low range instead of a single figure. The gap is checked by using a feeler gauge blade with the same high-low range. If the smaller end of the blade fits inside the gap, but the larger end is too big, the gap is okay. (See Figure 16-2.)

Feeler gauge blades (or wires) are calibrated in inches or metric units. Sometimes both types of units are marked; sometimes reference numbers are used instead of the actual thickness. For instance, a series of blades ranging in thickness from 0.001 in. to 0.009 in. might simply be marked 1 through 9. (See Figures 16-3 and 16-4.)

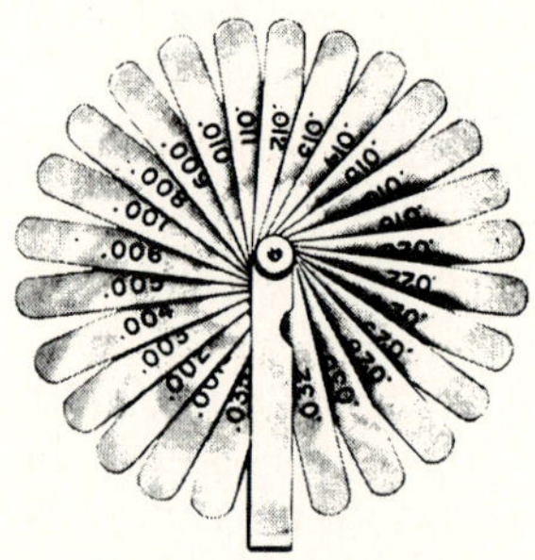

FIGURE 16-1. An all-purpose feeler gauge (Courtesy of New Britain Hand Tools).

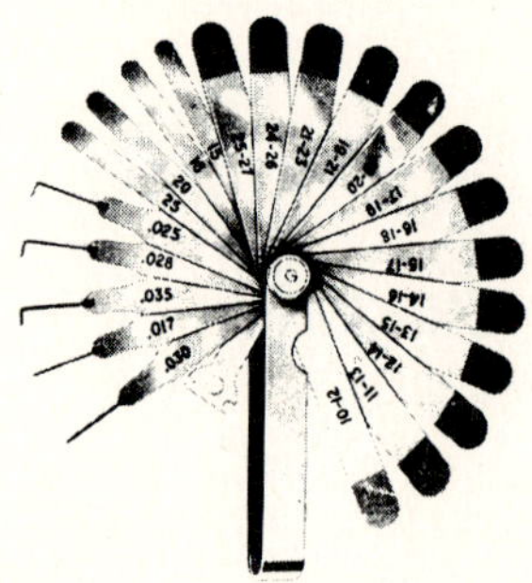

FIGURE 16-2. A wire and go-no-go feeler gauge set (Courtesy of New Britain Hand Tools).

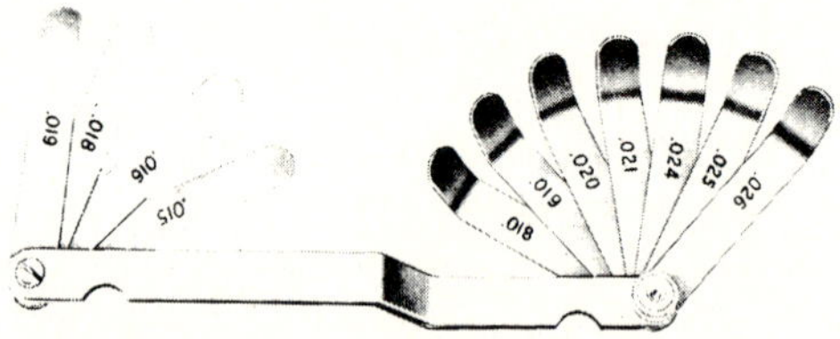

FIGURE 16-3. A combination overhead valve feeler gauge set (Courtesy of New Britain Hand Tools).

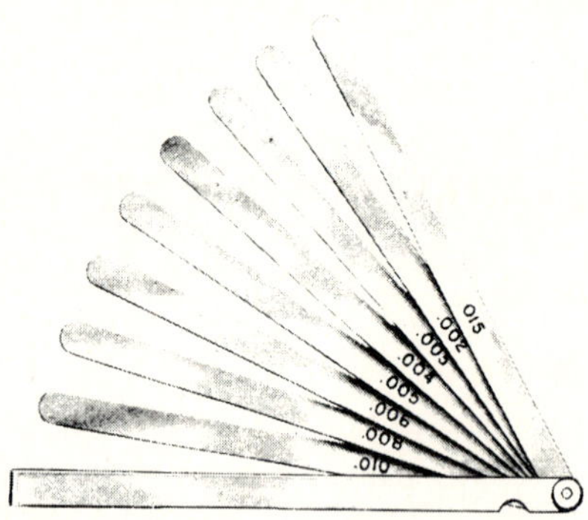

FIGURE 16-4. A piston feeler gauge set (Courtesy of New Britain Hand Tools).

CALIPERS

Two types of calipers are occasionally used by engine mechanics.

1. *Transfer type* calipers resemble well-made tongs with a thumbscrew for adjusting the position of the legs. (See Figure 16-5.) These calipers have no graduation and are not measuring devices as such. They are used to transfer an otherwise inaccessible dimension to a ruler for measurement. Although not generally believed to be accurate enough for most engine work, transfer calipers are handy for quickly comparing the dimensions of cylindrical objects (like pistons and cylinders).
2. *Vernier calipers* are precision instruments capable of measuring cylinder and shaft dimensions within a thousandth of an inch. (See Figure 16-6.) The average small engine technician does not use a vernier caliper as often as his counterpart in a machine shop.

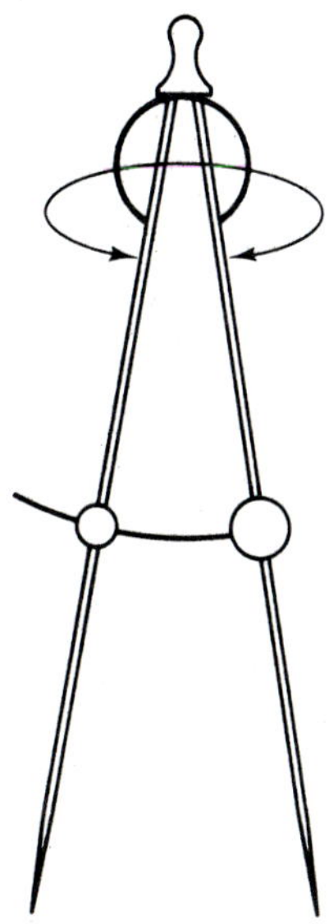

FIGURE 16-5. A caliper.

FIGURE 16-6. A Vernier caliper (Courtesy of New Britain Hand Tools).

MICROMETERS

Micrometers are more accurate than calipers, but they also cost a great deal more. Consequently, they are not found in every mechanic's tool box. However, most shops provide micrometers, since no quality measurements can be made without them. (See Figure 16-7.)

Micrometers contain two sets of graduations. Coarse readings appear on the sleeve body and fine readings on the thimble. After the micrometer has been adjusted so that the measuring portion fits whatever is being measured, the sleeve and thimble readings are added to get the total readings.

Micrometers use a vernier measuring system. Figure 16-8 and the following paragraphs describe how this system works.

1. Turn the thimble until the measuring part of the micrometer fits the span being measured. Never force the thimble; it can damage the micrometer.
2. As the thimble is being screwed in or out, notice how the edge of the thimble moves across the scale on the sleeve body. Every time the thimble

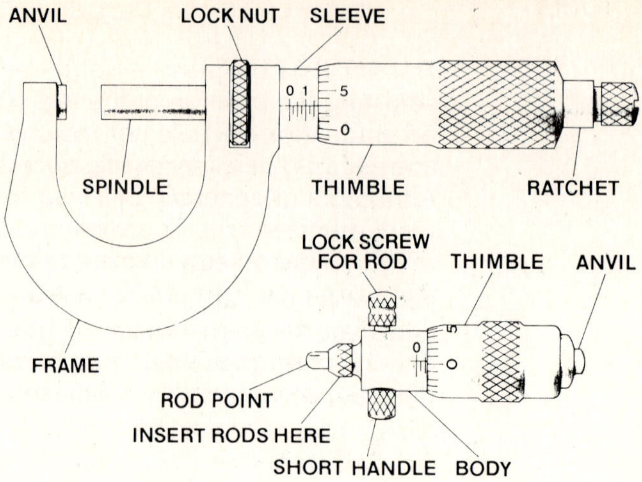

FIGURE 16-7. Outside micrometer and inside micrometer parts identification (Courtesy of Gould, Inc.).

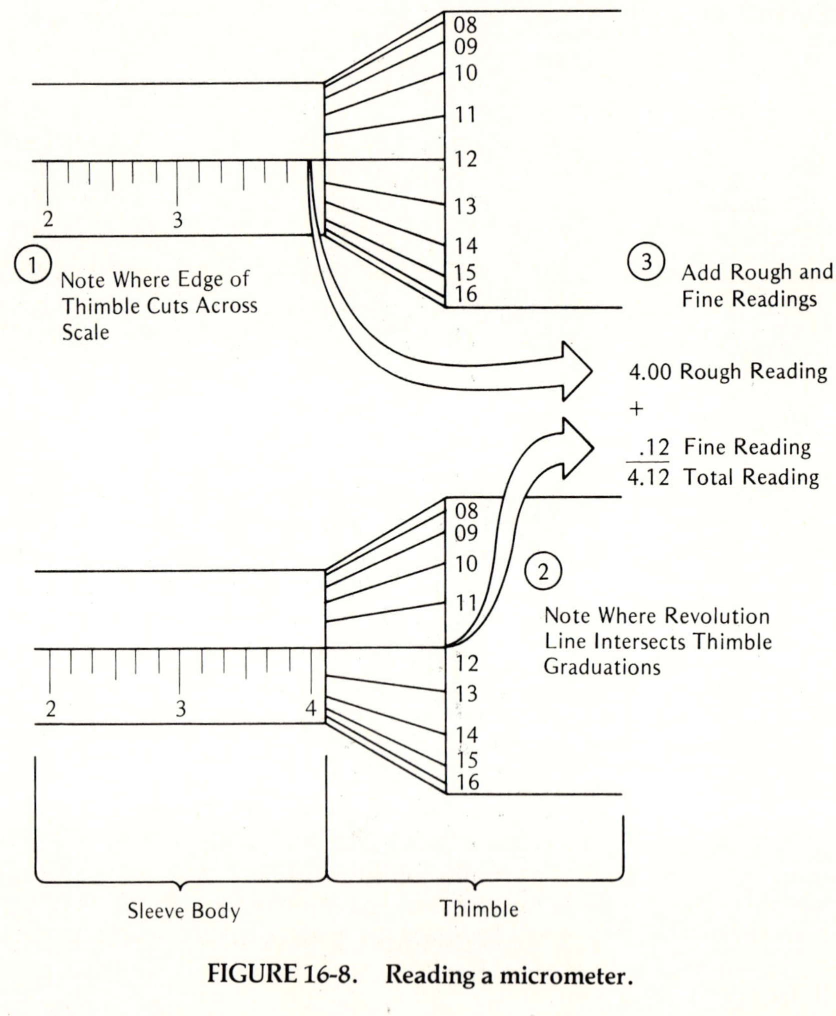

FIGURE 16-8. Reading a micrometer.

makes one revolution, its edge passes over one small division on the sleeve scale.

3. After the thimble adjustments have been completed, determine where the edge of the thimble cuts across the graduations on the sleeve. In the example shown in Figure 16-8, the thimble edge is nearest the 4.00-in. mark on the sleeve. Therefore, we can say that the coarse reading is 4.00 in. The object being measured is actually thicker because there is still some distance between the 4.00-in. mark and the edge of the thimble.
4. Observe the graduation marks on the thimble. They are used to subdivide the smallest graduations on the sleeve into even finer readings. In this example, the smallest thimble graduations are at 0.001-in. intervals.
5. Check to see where the horizontal (or revolution) line on the sleeve strikes the scale on the thimble. It is 0.12 in. This figure represents the fine reading; in other words, the distance from the 4.00-in. mark to the edge of the thimble. Adding 0.12 in. to 4.00 in. gives the total reading of 4.12 in., the actual thickness of the object being measured.

Figure 16-9 shows additional tips for reading micrometers. The best way to become proficient is to practice measuring objects of known thickness.

DIFFERENT KINDS OF LENGTH MEASURING SCALES

The three basic kinds of length measuring scales are fractional, decimal, and metric.

Fractional

Fractional scales divide inches into halves, quarters, sixteenths, thirty-seconds, etc. Most tape measures use fractional scales. So do most common household rulers.

Decimal

Decimal scales divide inches into tenths, hundredths, and thousandths. Many engine dimensions are given in decimal valves, so this scale is one of the most commonly needed. If a decimal ruler is not available, fractions can be converted to decimals by referring to a decimal equivalent table or by dividing the bottom part of the fraction into the upper part. For instance, to convert ⅛ in. to a decimal value, 8 is divided into 1. The result is 0.125 in.

Metric

Metric scales don't use inches at all. Distance is measured in meters, decimeters, centimeters, and millimeters. A meter, which is a little more than a yard, equals ten decimeters. A decimeter equals ten centimeters. A centimeter equals ten millimeters. All measurements in the metric system, including weight and volume measurements, are based on groupings of ten.

One hundred thousandths (.100″).

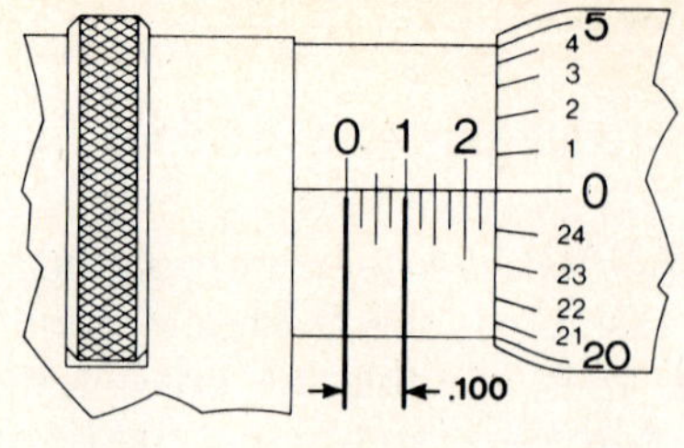

One thousandth of an inch (.001″).

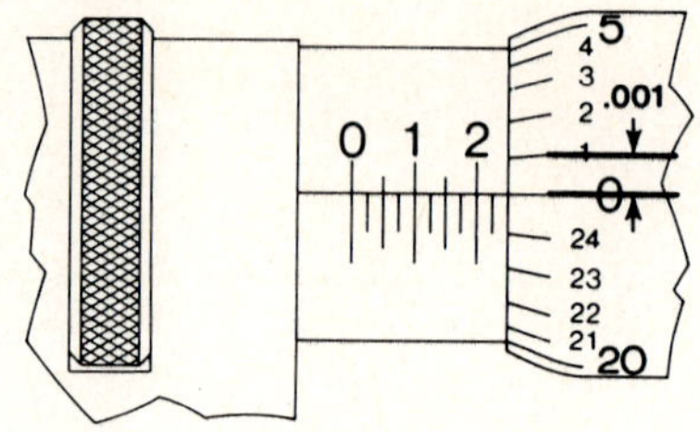

Twenty-five thousandths (.025″).

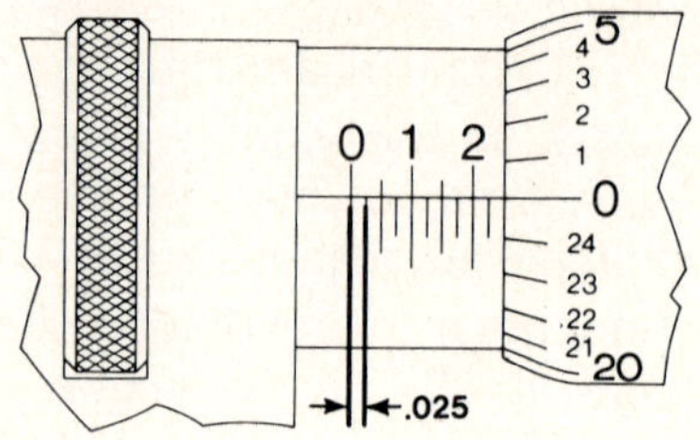

Estimating fractions of a thousandth.

HALF THOUSANDTHS

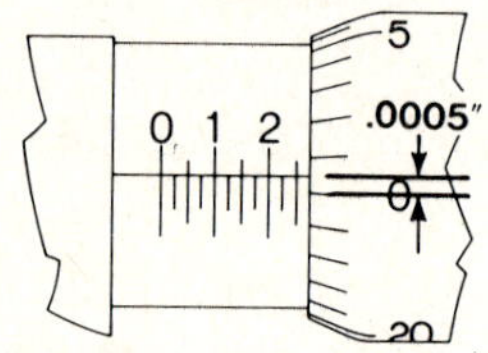

Actual Reading Shown .2755″

QUARTER THOUSANDTHS

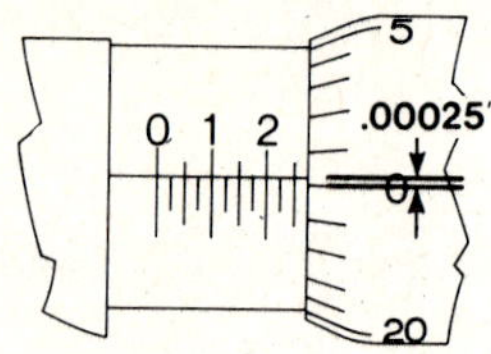

Actual Reading Shown .27525″

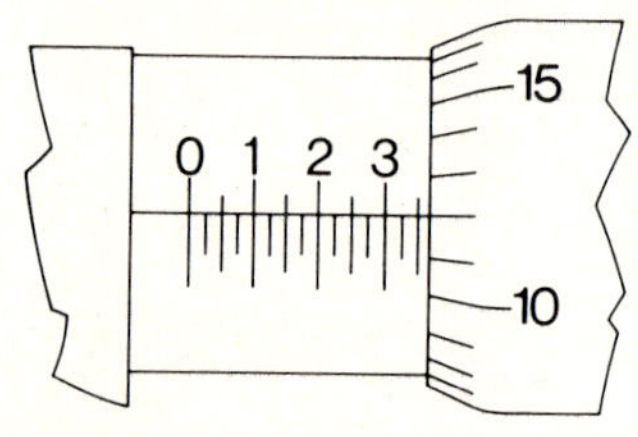

Micrometer set at .362″.

Five millimeters (5 mm).

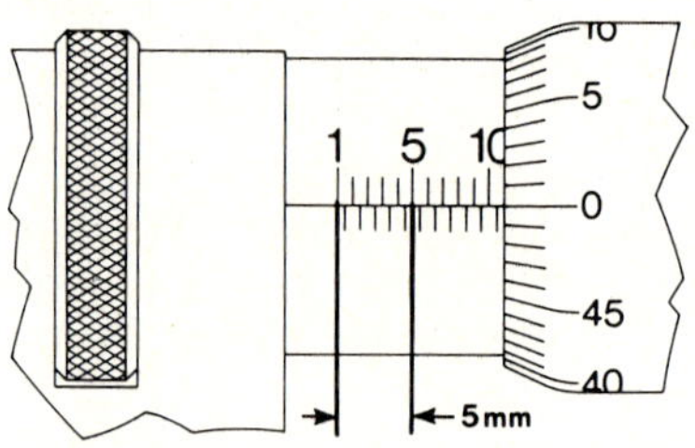

One hundredth of a millimeter (0.01 mm).

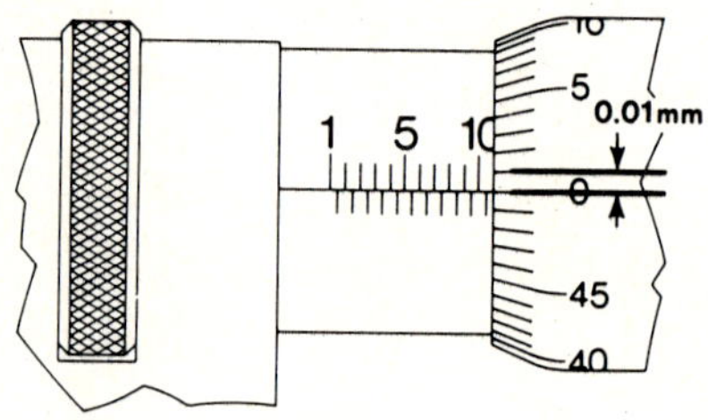

Five tenths of a millimeter (0.5 mm).

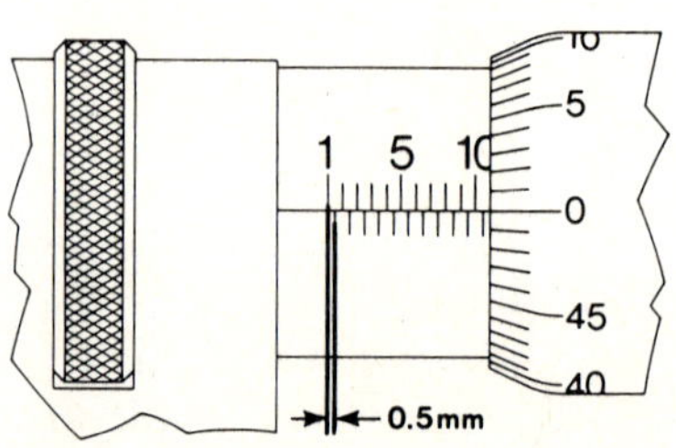

Micrometer set at 7.28 mm.

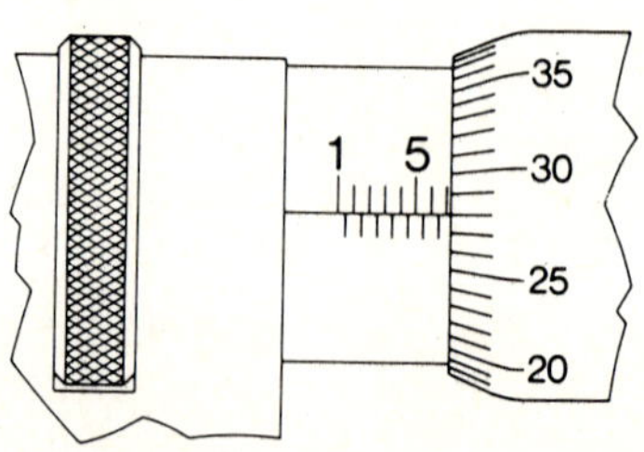

FIGURE 16-9. Additional micrometer tips (Courtesy of Gould, Inc.).

VOLUME MEASUREMENT

Besides needing to measure distances, mechanics often need to measure volume, e.g., the amount of liquid to pour into a container, the volume of a cylinder, etc. Common English or standard units of volume measurement are pints, quarts, gallons, and cubic inches. Common metric units of volume measurement are liters (approximately equal to quarts), milliliters, and cubic centimeters (ccs). Common measuring devices are graduated cylinders, flasks, and cups.

WEIGHT MEASUREMENT

It is also occasionally necessary to measure weights. Common weight measurement units are pounds, ounces, grams, kilograms, etc. Common weight measurement devices are scales and balances, both the portable and fixed position variety. Many shops have a weight scale of some kind.

TORQUE WRENCHES

Torque wrenches tell a mechanic how much force he is applying to a nut or bolt. The scale on the wrench is calibrated in pound-feet units. Most major service work should not even be attempted without a torque wrench (See Figure 16-10.)

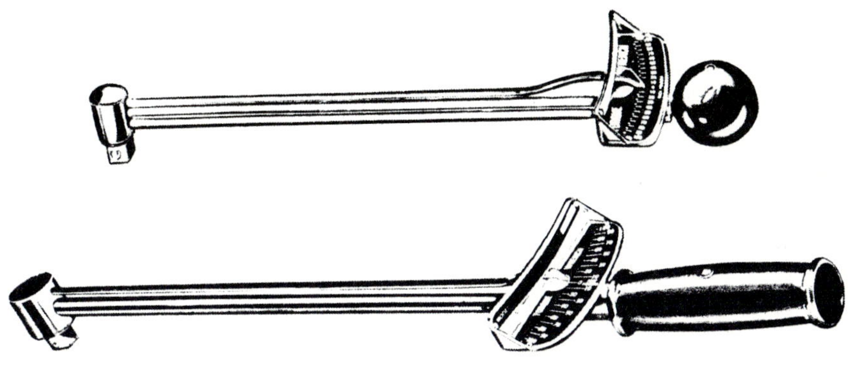

FIGURE 16-10. Torque wrenches (Courtesy of Gould, Inc.)

INDEX

D

E

F

G

H

I

K

L

M